IN VISIBLE LIGHT

Alfred Stieglitz. Equivalent, Music No. 1. Lake George, 1922 *(IMP/GEH)*

In Visible Light

Photography and the American Writer: 1840–1940

Carol Shloss

New York   Oxford

OXFORD UNIVERSITY PRESS

1987

Oxford University Press

Oxford   New York   Toronto
Delhi   Bombay   Calcutta   Madras   Karachi
Petaling Jaya   Singapore   Hong Kong   Tokyo
Nairobi   Dar es Salaam   Cape Town
Melbourne   Auckland

and associated companies in
Beirut   Berlin   Ibadan   Nicosia

Published by Oxford University Press, Inc.,
200 Madison Avenue, New York, New York 10016

Oxford is a registered trademark of Oxford University Press

Library of Congress Cataloging-in-Publication Data
Shloss, Carol.
In visible light.
Bibliography: p.   Includes index. 1. American fiction—History and criticism.
2. Literature and photography—United States. 3. Realism in literature.
I. Title.   II. Title: Photography and the American writer.
PS374.P43S5   1987   813'.009   86-18070
ISBN 0-19-503893-2 (alk. paper)

The lines from "The Tunnel" and "Atlantis" from *The Complete Poems and Selected Letters and
Prose of Hart Crane*, edited by Brom Weber, are reprinted by permission of Liveright
Publishing Corporation. Copyright 1933, 1958, 1968 by Liveright Publishing Corporation.

2 4 6 8 9 7 5 3 1

Printed in the United States of America
on acid-free paper

To my parents

ACKNOWLEDGMENTS

I would like to thank the scholars who have helped with this book and the friends who have sustained me while I was writing it. In particular I am grateful to Arlin Turner for a detailed reading of the Hawthorne chapter, C. E. Frazer Clark, Jr., and Rita Gollin for conversations about the dates of various Hawthorne photographs, and William Todd at the University of Texas, Austin, for sharing his knowledge of the Tauchnitz edition of Hawthorne's *Transformation*. Librarians at the Essex Institute, the New York Public Library, the Boston Atheneum, the Boston Public Library, the Massachusetts Historical Society, the Houghton Library and the library of the Fogg Art Museum at Harvard University gave me access to original materials. Portions of Chapters 4 and 5 appeared previously in *Boulevard* vol. 1, no. 1 (Spring 1986) and the *Virginia Quarterly Review* (Fall 1980), respectively. I am grateful for permission to reprint that material here.

I would like to thank Leon Edel for permission to quote from the letters of Henry James to Alvin Langdon Coburn, and the librarians at the Alderman Library, the University of Virginia, for making them available to me. The librarians at the Beinecke Library, Yale University, and in the rare book room at the Van Pelt Library of the University of Pennsylvania helped with materials about Theodore Dreiser and Alfred Stieglitz. Elinor Langer and Townsend Ludington shared their knowledge about the Spanish Civil War and helped me sort out the chronology of events in filming *The Spanish Earth*.

Emerson Ford in the interlibrary loan office at the Perkins Library, Duke University, was tireless in locating obscure books; Professor Irving B. Holley at Duke had excellent advice about the study of military history. Fred Pernell at the National Archives, Modern Military History Branch, helped me find unpublished records of the Signal Corps in World War II.

Wesleyan University provided me with a sabbatical leave; the National Endowment for the Humanities, the Bunting Institute of Radcliffe College, and the Rockefeller Foundation provided me with generous fellowships; Alex Harris and Bruce Payne invited me to be a fellow at the Center for Documentary Photography at Duke University. I am especially grateful to Alex, and to another friend, Bill Ravanesi, for helping me to understand documentary photography in its positive mode of "bearing witness."

I owe thanks to Martha Montgomery for giving me the opportunity to write as a Drexel Fellow in the Humanities and to Richard Ohmann, Joel Williamson, Samuel Cherniak, and Craig Eisendrath for their encouragement and reading the manuscript in its early stages. Ann Fasiola did an excellent job of typing. Finally I would like to thank Georges Borchardt and, at Oxford University Press, Sheldon Meyer and Stephanie Sakson-Ford. My friends, I trust, will recognize themselves.

Philadelphia   C.S.
December 1986

CONTENTS

IN VISIBLE LIGHT

We fondly supposed that we were representatives of mankind as a whole; and thus, all unknown to ourselves, we demonstrated our identity with the very privileged class we thought to repudiate.

—Simone de Beauvoir

[W]e are not talking about a gift that is bestowed but a power that is wrested from indifferent circumstances and shaped by unilateral desire.

—Max Kozloff

Most of the photographer types were wearing ... their light meters as pendants around their necks. If it was an art, it was the only one in which the artist wore something that made him visibly a practitioner.

—Paul Theroux

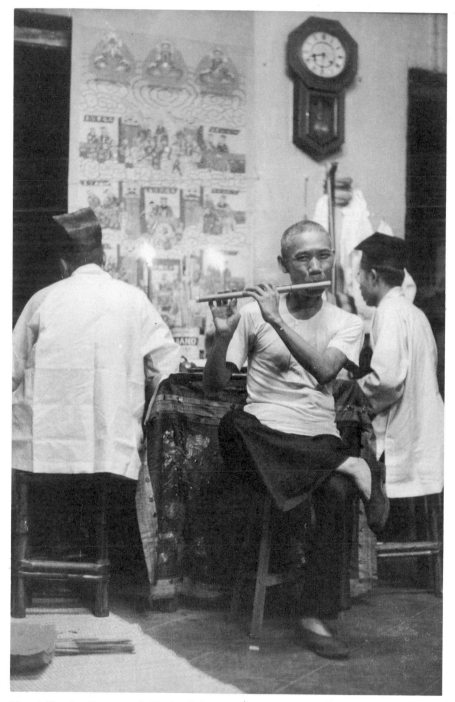

Henri Cartier-Bresson. A Taoist Priest Performing Funeral Music in the Death House. Singapore, 1949.

Introduction

FRAMING

THERE is a wonderful anecdote left to us by one of the first men to experiment with a camera. Still excited by the photographic lecture he had just attended, Marc-Antoine Gaudin rushed around Paris gathering bits and pieces of stuff to make the appropriate apparatus, and then set the miraculous procedure in motion:

> As soon as day broke . . . my camera was ready. It consisted of three-inches focal length, fixed in a cardboard box. After having iodized the plate while holding it in my hand, I put it in my box which I pointed out of my window, and bravely waited out the fifteen minutes that the exposure required; then I treated my plate . . . with mercury, heating the glue pot which contained it with a candle. The mystery was done at once: I had Prussian blue sky, houses black as ink, but the window frame was perfect![1]

Gaudin's dismay was understandable, his self-possession admirable. He saw himself as the butt of a good joke, rather like the living moral to a fable where vanity is foiled and expectations shown to be foolish. In

3

taking a picture of his window frame, he had made an indelible record of his own mistake.

It is a good anecdote about trial and error; but to me, there has always been something more to this vignette. For what Gaudin had photographed was, after all, a frame—which surrounded and limited his experiment, which spoke to him of enclosure and separation, and which showed his art to be an art of coming upon, an *ars inveniendi*. Had he succeeded by his own standards, the frame would have been invisible, represented only by an absence or by the openness of white space. But Gaudin's mistake is more interesting than is his lost achievement, for in picturing a line of demarcation in this manner, in noticing a border, a boundary, an edge, Gaudin unwittingly encoded his own exclusion, his own position on the other side, and his own subsequent need to transgress distance, to approach. A picture of a frame is a picture of what art usually leaves implicit about its own status as something discrete from what is outside or around it. It is also a picture which calls for completion since it asks us to visualize a relation to the frame, a way of coming upon it or of entering into it. Tolstoy thought the function of art to be the destruction of separateness; he wanted to educate people to unity. But unity is at most a longing or a goal; and Gaudin's window frame reminds us that, whatever its desired end, art is always drawn out of separateness and that the artist's terms of engagement with his or her subject are manifold, with referents in fear, hope, desire, fascination, and hunger.

COMING UPON

This is a book about how artists negotiate approaches to their subjects. Like the story about Gaudin's window frame, it is a meditation about what surrounds a work of art, or, to use the words of Fredric Jameson, it is about the "empirical preconditions that must have been secured" in order for a narrative or an image to have assumed a particular shape.[2] There are many ways to come upon a subject, many ways to maneuver through the circumstances that initially nourish creativity. The important thing is to notice these circumstances and to reflect on all that is contained within them.

One of our contemporaries, John Berger, a London-born writer who is living and working among the French peasantry, has done just this: he has spoken about the complexities of his own position among strangers whose archaic traditions will provide the basis for a three-part novel, *Into Their Labours*. As he put it, "Whatever the motives, political or personal, which have led me to undertake to write something, the writing becomes ... a struggle to give meaning to experience. Every profession has limits to its competence, but also its own territory. Writing, as I know it, has no territory of its own. The act of writing is nothing except the act of approaching the experience written about."[3]

What qualities could guide this cartography, this mapping of the spirit? What distances between people needed to be finessed? What distinctions to be made? By choosing to share the discomforts of peasant housing, manual work, and village ceremonies, he decided, he was bound to the subjects of his work by circumstance. In that he was a parent bringing up a child, he shared a structure of family life. But though the same weather buffeted them all, it could not eclipse those things that set him apart, that defined him in terms of a different religion, patrimony, kinship, and economic prospect. It could not obscure what was to me, and I think to him, the most significant distinction between them—his lack of history in one place. By virtue of moving there from somewhere else, he had a perspective, a distance, a catholic measure of experience that was denied to the peasantry and became, in turn, the subject of many of Berger's own reflections. Suspended between self-knowledge and lack of panoramic knowledge, aware but unable to measure, longing but without means, Berger's peasants were threatened both by industrial capitalism and by the inability to see themselves in relation to that threat. At most they were like Pépé in "The Wind Howls Too," whose outermost achievement was the ability to recognize the possibility of another more encompassing perspective than his own: "I'd look down at them like the old crow looks at us!" "That is what I would like to know if I was a crow on a tree watching."[4]

Berger knew that he was not writing about people who had no stories about themselves. Speech, in itself, was not the issue, for, as he could see, peasants habitually gossiped in order to reflect upon the meaning of their lives. "Without such a portrait ... the village would have been forced to doubt its own existence."[5] But it was impossible for him not to notice that this portrait had no transcendence, no overview, no balanced sense that life can have another form. Pépé's inability to reflect on himself from the point of view of an observer, to see himself like "a crow on a tree watching" his own activity, meant that he could not control his life in certain ways.

Understanding these things placed Berger in the most delicate of positions. Because witnessing can validate a fragile sense of worth, his presence in the village was a virtue. Because watching threatens to take a depository of meaning out of the hands of those who have generated that meaning, he was also a potential danger. Whose version of the truth would prevail? Berger proceeded as if he understood his work to be a kind of human commerce or a form of cultural exchange—one country with another—which had its own strategies for knowing and being known, for gaining access to the world and being permitted or denied that access. Like any form of commerce, Berger would have us understand, the experience of making art can be mismanaged or negotiated with grace; it can be treated as a kind of theft which inspires a responsive secretiveness or as a mode of bearing witness which is itself a valuable service.

I have spoken at length about John Berger, but the tension he experienced in France, generated between the poles of violation and generosity, between the extremes of distortion and truth, can disturb the work of any artist who takes the issue of cultural dispossession seriously. Theodor Adorno, the German social philosopher, has said, "The need to lend a voice to suffering is a condition of all truth."[6] But lending that voice, presuming to speak truthfully for those one can only approach, who are framed and separated by perspective and heritage, presupposes the resolution of an entire set of power relationships; it presupposes a decision about a posture, a stance, a style of being present in the world of those who are considered voiceless.

APPROACHING WITH THE CAMERA

Usually the resolution of these choices—decisions about handling the circumstances that I have called the empirical preconditions of art—is implied. In photography it is implied but it also bears direct relation to an actual moment when the photographer stands posed before his subject, the subject posed before him, the frame a technical feature of the camera itself. It is this circumstance that places photography as the starting point of this book. Without a photographer's presence, without his trespassing glance, his engagement, no images would exist: we can see photographers at work, we can ask questions about their procedures, their ways of traversing distance, and we can use their experiences as analogues for the more hidden activities of writers who are gathering materials. Through watching them, we can trace the ways that contingent historical circumstances are linked with questions of aesthetic figure; we can understand how art grows out of experience, and how it transposes these originating moments into forms which continue to be haunted, empowered, or sometimes subverted by them. It is as if, by this watchfulness, we can gain a stealthy vantage point which allows us briefly to violate Plato's assertion that that which enables us to see must itself remain invisible. For the camera's literal, dependent presence before a subject lets us draw visibility and the ways of organizing visibility into another kind of light where we can see the nuances of negotiating with the external sources of one's creativity.

Henri Cartier-Bresson has spoken repeatedly about these instants, these moments of collusion when the world is delicately and perfectly joined with an interior predisposition. For him, the "decisive moment" was a moment of complete reciprocity, the "simultaneous recognition in a fraction of a second," of the unity of photographer and subject, the congruence of self-expression and revelation of the Other. "[T]hrough the act of living," he said, "the discovery of oneself is made concurrently with the discovery of the world around us."[7] According to him, the photographer's struggle was always to arrive at these fugitive balances, to be unobtrusive, but close, to find a mode of receptivity and recognition that was also a revelation. Lines, surfaces, values in a print furnished a vocabulary of the spirit made concrete; they were a testament to another, unseen harmony.

Of those who followed him, Walker Evans paid clearest tribute to the value of Cartier-Bresson's "simple factual testimony." The French photographer spoke of his own need to have "a velvet hand, a hawk's eye," and of the value of unobtrusive approaches; Evans approached in these ways—smoothly, astutely, without intrusion, as if the camera were only lending a surreptitious visibility to the world. If a quiet presence can be the sign of reverence, then Evans was a novitiate of the given, a man whose homage was measured by his own self-effacement until James Agee recognized the enormity of his talent and lauded him, as well as the Alabama tenant farmers Evans photographed, in *Let Us Now Praise Famous Men*.

Agee admired Evans, but he himself struggled more vehemently with the social implications of the photographic moment than did his colleague, seeing its potential for spilling over into damage or unwitting manipulation. Where Evans went on to use a hidden camera in the New York subways, taking the impersonality of the recording machine to its limit, Agee remained tortured by the possible misuses of art, by the problematic nature of looking at strangers who were joined to him only because curiosity inserted him in their midst. When he and Evans tried to photograph a small, country church in Alabama in the late 1930s, Agee winced at their joint audacity. Evans had set up a tripod and was waiting for the direction of sunlight to change when a young, Negro couple walked by. Their presence changed the occasion and made the two observers "ashamed and insecure in our wish to . . .

possess their church." Knowing that Evans would take the photograph in any case, Agee nonetheless ran after the couple to secure their permission, scared them badly, and then regarded his behavior as a "crime against nature."[8] Behind this heavy, self-accusing rhetoric was a finely honed regard for the vulnerability of the blacks, people for whom investigative reporters represented power and possible damage. In his eyes, images could either be "offered" or "stolen," and while one term implies the subject's consent and the other his non-compliance, both words show that Agee considered photography—and, by extension, his own research—as a kind of displaced monetary transaction, or to use a phrase of Susan Sontag's, an example of "consciousness in its acquisitive mood."[9] He thought it no accident that pictures were usually taken by one class of another, that middle-class people usually took pictures of the undefended poor, that "concerned" photography was usually a dubiously motivated concern for those who could most easily be taken from. He put it succinctly: "Art and the imagination are capable of being harmful."[10]

If, in retrospect, Agee's caution seems exorbitant, it nonetheless alerts us to the possibility of other, less classical modes of approach than those practiced by Cartier-Bresson and Evans. Agee wanted to emulate the "security and satisfaction and patience"[11] of Evans's temper, but he understood such poise to be a rare quality; he knew that desire or loathing could undermine the stasis advocated by Cartier-Bresson, that kinetic emotions—hunger, depletion, or greed—were as common in the creation of art as the attentive balance he desired for himself. Decisive moments, Agee would have us understand, can express more than equilibrium.

Some photographers make love to the world with their cameras. "To be with Miriam was a fulfillment," Edward Weston wrote in his daybooks. "I retain impressions. . . . I find myself climbing once more to her hill-top, or racing with her over the white sand of Carmel . . . or pointing my camera towards her naked body."[12] Alfred Stieglitz's famed, erotic monument to Georgia O'Keeffe is another example that comes immediately to mind, for it was a collaboration which, as Janet Malcolm points out, broke Dorothy Norman's guileless portrait of Stieglitz as a "pompous, sententious, petulant, cold, sexless old maid"[13] and allowed him to emerge with a dark, provocative sexual identity of his own. No one lacking a secure masculinity could have evoked those secret glances, those expressions of waste, of energy spent, of sadness. Their loveliness gave rise to an occasion both Stieglitz and O'Keeffe were fond of recalling, an afternoon when other men at the Anderson Gallery wanted Stieglitz to photograph their women in a similar mode.

"If they had known," O'Keeffe said, "what a close relationship he would have needed to have to photograph their wives or girlfriends the way he photographed me—I think they wouldn't have been interested."[14]

For both of these photographers, the camera was used as an extension of intimacy, a visual caress. They integrated making images with making love, seasoning their work with a strange boastfulness that both claimed the women they portrayed and snubbed the viewers who bore witness to a collusion that would forever exclude them. If Weston's or Stieglitz's subjects were aroused, it was because they were aroused by them, showing in the most blatant way possible what Robert Frank has claimed to be more generally true: "[I]t is always the instantaneous reaction to oneself that produces a photograph."[15] Decisive moments such as these are informed by an exclusive, but mutual looking back and forth.

If desire characterizes the approach of some photographers, then so does greed or fascination with the misshapen, the odd, the unfortunate. To some extent, Georgia O'Keeffe and Miriam, Weston's lover, invited the camera's scrutiny; their nakedness was an offering. What of photographers like Don McCullin, whose images of war and suffering recorded the intersection of the public and the private, the place where policy decisions about combat or international trade or civic priorities take their toll on the bodies of individual civilians? John Le Carré, who visited McCullin around 1980, tried to get at the root of his addiction to mutilation and death, to understand the drive to collect such a "kitbag of horrors." Could one feel, he asked, a nostalgia for pain and lost poverty? Was regret limited only to the loss of love? Could the moment of connection with one's subject, the "naked affinity," the "yes," be as strong in the face of mutilation as in the presence of eros?

> The moment ... when he or she sees him, and forgives him, at death's edge, starving, inconsolably bereaved, when their own child lies dead on the hall floor, bombed in the attack: still "yes." Yes take me. Yes take us. Yes show the world my pain.[16]

To Le Carré, McCullin's bonding was a bonding in misery, a connection of inner pain with the tangible correlatives of pain provided by the world's famines and disasters. Because of this recognition of the self in the Other, McCullin's photographic approach can also be considered a variation of Cartier-Bresson's decisive moment; it is a commerce between equals—although those who trade deal only in deprivation and death. McCullin's connection to the source of his subjects' suffering prevents his photography from becoming a form of trespass.

Brutal, jarring, lacking the still patience of Cartier-Bresson's images, they nonetheless show more affinity to the classic French work than—to take another example—to Diane Arbus's photographs of American freaks, nudists, and middle-class pariahs, which they resemble in more overt ways.

The distinction is measured by the concept of reciprocity or mutual recognition. Where McCullin could come upon pain because he already knew it, Arbus looked to her subjects to provide escape from the self, to offer a dimension not already present in the psyche. "My favorite thing," she said, "is to go where I've never been."[17]

Her photography became a strategy for handling the isolation and, as she saw it, the protective numbness of middle American experience. Haunted by her own lack of trauma, in search of the ineffable marks of experience, she chose as her subjects those whose deformity announced a contrary world, a life of combat in that world, a triumph of survival. "They're aristocrats," she said of humpbacked giants, midgets, and morons. They had met the adversity she had never felt as a child; they were "real" in a way she did not feel herself to be real.

The movement between Arbus and her subjects is, then, an imbalanced one, a going-out-into the unrecognizable world from a position of depletion or need. To Arbus, the camera was a "power thing." There was an edge of advantage to it, a license to intervene. It was a silent but yearning voice; the subject served as its echo, confirming by the simple existence of a response, the existence of an originating movement. It both answered a cry and careened out into the unknown. An anecdote Arbus told serves equally to explain her aggression in the face of personal inadequacy: she is like the man in the photograph which a whore once handed to her: he is sitting on a bed in a borrowed bra, "like anybody would try on what the other person had that he didn't have."[18]

It was this trying-on of alien experience that marked Arbus's photography as both damaging and suspect, an example of the transgression James Agee so feared in his work in the rural South. When she wrote about her own efforts, Arbus stressed the identification she established with her subjects. But in what could this have consisted, since, in her own words, their deformity was a badge of experience that life had denied her and would always deny her? The only identification possible was through her own voyeurism, made legitimate by donning the camera to create a purpose for it. Arbus's pictures are disturbing not only because of the grotesqueness of the subjects, but because of the posture that implicitly accounts for all the surface permutations: the photographer's hunger for the grotesque, and her anxiety in its absence. If Franz Kafka's hunger artist ("Der Hungerkunstler") made an art of star-

INTRODUCTION   11

vation, turning his own hunger for an unobtainable spiritual nourish-
ment into a public spectacle, then Arbus works a strange variation on
the theme, displaying both the urge and its satiation, a psychic gluttony
and a stilled craving. Such taking implies a cost.

Whenever reciprocity is violated in these ways, whenever there is
a dissymmetry in trade, it becomes appropriate to ask questions about
the exercise of power. A method of observation need not be a power
tactic, but it is capable of being one. It can, as it does in this case, require
the submission of the subject to the purposes of another—a violation of
will that inheres in the activity of photographing itself—or it can,
because the image is the expression of an alien sensibility, suppress the
subject's countervailing view of what is the case. It can make him or
her an object of knowledge rather than a subject in communication; it
can usurp the generation of meaning. Do the elderly people who have
been crowned king and queen of the dance (Arbus, "The King and
Queen of a Senior Citizens Dance, N.Y.C., 1970") see themselves as
ugly and ridiculous in their imitation grandeur? Do the young teenagers (Arbus, "Teenage Couple on Hudson Street, N.Y.C., 1963") know
that their clothing is badly cut and too small for them? Do they recognize that their awkward embrace parodies adult romance? Do they
know that their picture acquires meaning for Arbus as part of an
inquiry into the nature of the grotesque? These are the kinds of problems that John Berger worried about as he worked among the French
peasants, and they are particularly hard to address because such discrepancies of will and cognition can exist apart from the artist's intention
to subjugate or the subject's knowledge of being manipulated. They can
arise out of the structure of negotiation itself. Arbus's teenagers may
have agreed to pose without knowing what they were posing for; they
may have consented in ignorance.

In these several modes of approaching the subject, we can see cor-
respondences to other ways of organizing visibility in the political
world; those behaviors I have called photographic "commerce" are
actually modes of human interaction that are capable of analysis in
wider terms. Cartier-Bresson and Evans, with their stillness, balance,
and respect for the thing-in-itself, Stieglitz and Weston with their erot-
icism, their movement toward, their evocation of an answering respon-
siveness, McCullin within his closed circle of common pain: all establish momentary, elusive, but concrete instances of investigation that
admit the participation of the subject. Behind these ways of working,
so various in formal results, lies what we might call an Aristotelian
concept of political action. It is a view which influenced Rousseau,
Hegel, and Marx, and it places primary value on men's and women's

efforts to maintain the power of self-reflection and to control their collective affairs. As Rousseau put it, "[W]hat generalizes the will is not
so much the number of voices as the common interest which unites
them."[19] In the words of a contemporary social philosopher, Brian Fay,
"[W]hat is fundamental about [the Aristotelian vision of] politics is ...
the involvement of the citizens in the process of determining their own
collective identity."[20] Whatever else may be said about these photographers, their work expresses a vision of what the social world could
become were such "decisive moments" generalizable. It recognizes the
importance of the ideas men and women have of themselves, of what
they consider "appropriate, right and fitting," and what they feel they
are worth. It is inevitably a collusion, that is, it is a form of interaction
in which meaning is generated collectively. And it is this quality that
John Berger recognized in the photographs of Paul Strand when he
lauded his "ability to invite the narrative: to present himself to his subject in such a way that the subject is willing to say: *I am as you see me.*"[21]

Behind Diane Arbus's work lies an alternative model which proceeds from the belief that there is, and should be, some way of privileging a single viewpoint—whether through raw power, strategic
advantage, or the superior claims of expert knowledge. If we were to
think about this model in the contemporary political realm, we would
have to imagine a society governed by the decisions of technical experts
whose special knowledge qualified them to make judgments on behalf
of those who are untrained. They would claim to be both knowledge-
able and ethically neutral, but their assertions of objectivity would, in
fact, cloak a submerged interest which would place them in the figurative position of rulers surveying their territories, armed not with
armies that implement a literal power, but with knowledge that effects
the same structural result: a society that creates an elite of qualified
experts creates a hierarchy just as potent as that based on a more literal
force since it is not possible for non-literate, that is, non-expert, citizens
to challenge the validity of decisions made on their behalf.[22] In either
case, there can be no structure of accountability, no dialogue, no mutual
involvement in establishing identity. The subjects of knowledge cannot
be self-reflective, but become, instead, dependent upon a hypostatized
or seemingly anonymous power. The looking-back-and-forth of the
Aristotelian model is broken; visibility is organized in one direction
alone as those-who-see implement policy and those-who-are-looked-
upon bear its consequences. Walker Evans had judged American life to
be organized according to this structural model when he introduced his
New York subway photographs by saying, "As it happens, you don't
see among them the face of a judge or a senator or a bank president.

What you do see is at once sobering, startling and obvious: these are the ladies and gentlemen of the jury."[23] These are the citizens in whose faces and in whose tired and listless bearing you can read the judgment of the dispossessed.

In its most sinister form, this organization of social visibility was expressed in Jeremy Bentham's panopticon. In his prison, which served as a distilled and heinous figure of a social order which he wanted to make more general, visibility was never mutual, it was in a radical state of disequilibrium whose goal was the enforced behavior of those who were observed. The architecture of the prison served its social aims perfectly: a central observation tower hid authority from sight, while the open-barred cells surrounding it offered the prisoners to view. The disparity of visibility both expressed a disparity of power and served to perpetuate it, since the fear of observation and the knowledge of its possible, constant presence were as effective in enforcing conduct as an actual, perpetual surveillance. Because an inmate of Bentham's structure could see neither the presence nor the absence of authority, he or she could never experience even temporary freedom. The possibility of the gaze fostered self-scrutiny; it engendered behavior that seemed voluntary but was not. Michel Foucault suggests that Bentham's prison can serve as the emblem of any society in disequilibrium, for this kind of organization of visibility inevitably expresses a relationship of force.[24] When the source of coercion remains inscrutable, it is incapable of being addressed: one cannot "be" with that which does not appear to be there; one cannot even discern the nature of one's strategic disadvantage. Thus there can be no struggle, no resolution, no restitution of balance.

To associate the work of Diane Arbus with this paradigm is not to claim that she is a technical expert, for her photographs neither claim neutrality nor effect literal control of any person or policy. But it is to press the view that her way of approaching the moment of photographic engagement organizes society into a momentary hierarchy which the print perpetuates in time. Her approach is privileged in that its motive remains hidden from the subjects of her observation and thus outside the realm of dialogue; it is privileged in that the meaning of the encounter remains in her hands. With her hunger for foreign experience, with her excursions into places of psychic damage, she reminds us that it is possible to engage in a kind of panopticon of the spirit, where observation, no matter what its overt purpose, serves to gain an advantage for the observer at the expense of the subject. To treat a subject in instrumental terms, to think of him or her as a useful means of realizing a goal that is covert and extrinsic to the immediate transaction, to prohibit reciprocal interaction is to assert a privilege of perception which replicates the political domination effected by Bentham's hidden-observation tower. As Arbus said, her photography pressed an advantage; it was a "power thing"; it placed her in the position of an arbiter or judge even though her judgments often took the form of oblique admiration.

In calling attention to these two models—one that acknowledges the autonomy of the Other and another that submerges the Other in a solitary purpose, one that defines itself by including the self-knowledge of the subject and another that disqualifies meanings generated externally—I am identifying extremes. But I think that all photography as observation of the social world plays against these two poles, that it violates or reinforces or edges out toward issues involving human freedom, the exercise of the will, and the status of knowledge. In this sense, to approach the territory of another—even though that approach is made through the viewfinder—is to approach individual property and to indicate through that approach a more general vision of human polity.

APPROACHING WITH WORDS

Why should we choose photography as an interpretive code through which to read and receive written texts? Why should we gloss one kind of creativity with another? If all writers were like Samuel Beckett, whose project was a solitary gesture against despair, against hope, against the solitude of creation, and equally against its dependence upon others, the analogy would not suggest itself. Writing would refuse the world, would articulate only one's own pain, one's own confinement. But if one has the predisposition "to seek out . . . others," to experience the world as unknown and unexpected and to create out of that opening-up-to the presence of another, then photography is, perhaps, an inevitable paradigm. Any writer who gathers the world into his or her imagination, who is, in the words of Henry James, a "seeking fabulist,"[25] knows that he or she acts in hidden but analogous ways: constituting the text, finding its material, is similar to using a camera.

This is, I think, what struck James Agee so forcibly when he worked with Walker Evans in the American South during the 1930s. This is, at least, what struck me about James Agee: he was curiously but profoundly worried about working with a photographer. His editors at *Fortune* magazine may have sent Evans down to Alabama to provide illustrations for Agee's more commanding text; but for Agee, the *photographs* were the commanding text. Far from being an added dimension to a story, reinforcement or proof of some other mode of representation, they provided the light by which Agee understood his own creativity, his own humanity and his lack of it. "Next to unassisted and weaponless consciousness," he said, "[the camera is] the central instrument of our time."[26] It was as if he saw in those several dazzling and uncomfortable weeks of watching Evans a visible tableau of his own position in the world, the hidden and secretive probing of the writer made tangible, its effects on others exposed. Evans's methods of working with his subjects suggested to Agee that a once actual way of handling authority could serve as a key to understanding the verbal forms of author-ity, that a model of negotiation with the world underlay the text and the postures of its speaking voice.

It is a stunning reversal of habitual attitudes; and, in a sense, this book is a prolonged meditation on the meaning of that reversal, an inquiry into its origins in American history and its permutations through time. Once seen, it is impossible not to continue to see that one of the most pressing problems in our literary history is the problem revealed to us by the camera: the problem of coming-upon, of approach, of the politics enacted in and through art—not as something that precedes creativity or that stands to the side of it, but as something enacted through the creation of a text and something that remains embodied in it.

We can see this preoccupation with the conduct of art as a subtext in the work of Nathaniel Hawthorne. In retrospect it does not seem fortuitous that he was the first of our writers to think about daguerreotypy as a mode of social investigation or that *The House of the Seven Gables* used photography to explore the implicit contract established between an artist and his subjects. "I misuse heaven's blessed sunshine by tracing out human features through its agency," Holgrave said in uneasy defense of his images, identifying both the power and the trespass that Hawthorne thought to be inherent in "sun-writing" as well as in his own more clandestine ways of coming upon the world. He considered his subjects, like Hepzibah Pyncheon, to be "tortured ... with a sense of overwhelming shame that strange and unloving eyes should have the privilege of gazing"; but he persisted in watching them because his desire for truth continued to override his fear of violating privacy: "There is a wonderful insight in heaven's broad and simple sunshine," Holgrave continued. "While we give it credit only for depicting the merest surface, it actually brings out the secret character with a truth that no painter would ever venture upon."[27] Photography allowed Hawthorne to imagine an art that was true because it was free from selection and bias, and it allowed him to imagine an art that was blameless for the same reason. If the camera were autotelic—that is, if the sunlight rather than a human sensibility were responsible for the image—then it was possible to resolve the problem of authorial mediation which he found so deeply troubling and which caused him repeatedly to wish for invisibility so that he might seize the social world unawares.

The difficulty of achieving such clarity, of standing in a neutral place, became the submerged subject of Henry James's fiction as well. He shared Hawthorne's troubled sense of the liability or, as he put it, "the expense of vision." For him, the making of art was a consistently subtle negotiation which could never be resolved by fantasies about personal transparency. Though Hawthorne desired it and the French realists claimed to have practiced it, James considered autotelic form a myth. His belief in the inevitability of subjective appraisal in art formed the substance of his quarrel with the French realists, and his novels almost ritualistically reenacted the cost of attempting to cloak social inquisitiveness and personal acquisitiveness in the guise of impartiality.

Although James pointedly borrowed the language of photography to explain the dynamics of narrative (it was he who most actively pressed the metaphors of lenses, frames, and points of view into service to show us the pressure of one discipline on another), he did so with remarkable irony. Where Zola and his associates had spoken of realism as a studied objectivity before the subject, a writing-as-if-one-were-not-present which only the camera could describe, James withdrew both photography and narrative writing from their association with neutrality of observation. It was he who anticipated later criticisms of positivism and showed the concealed interest (point of view) that lay behind supposedly value-free investigations. He thought Zola's attitude toward photography to be an example of the institutionalization of scientific discourse and wrote in implicit rebuke to it. In retrospect one can read James's narrative strategy as a submerged but continuous response to Foucault's question: "What types of knowledge do you want to disqualify in the very instant of your demand: 'Is it a science?' "[28] Zola's demand to writers: "Is it what the camera would record?" was, to James, part of science's unacknowledged struggle to subjugate the self-knowledge of the subject; it constituted an implicit statement about the nature of authority and the eclipsed posture of the subject in the presence of authority. If James could not conjure an alternative model of discourse, he could at the least articulate the treachery involved in his practice of art; he could dramatize the discord generated by denying the multiplicity of knowledges that constitute a balanced culture.

And so it has continued through time. Hawthorne and James were succeeded by other writers who brooded on the meaning of photography. But whatever qualities of temperament or accidents of history made each of them unique, they were bound by a common concern about creativity as a mode of action in the world, a concern that was honed on images and refined by example. If Henry James considered his fictional narrators to be "convenient substitute[s] or apologist[s] for the creative power otherwise so veiled and disembodied,"[29] then we might say that photographers have served writers as narrators of the actual, embodying creative powers in the most literal ways, unveiling the dynamics of vision, insisting that methods of working, ways of gaining visual access to the world, of relating to subjects, constitute a dimension of art as substantive as—indeed consubstantive with—the more formal qualities that commonly define an aesthetic tradition. It was as if they confirmed in action, in gentures of real consequences, what William Carlos Williams could affirm only in words, that "to write nine tenths of the problem/is to live."[30]

For the American realists who came after Henry James, these problems of living and writing occurred as part of a debate about the uses of disinterest in the creation of art. As a group, they tended to be preoccupied with issues of detachment and truth and thus to consider the photograph as a "window" onto experience. Emile Zola articulated one of the primary presuppositions of the movement when he said, "The observer sets down purely and simply the phenomena he has before his eyes. . . . He ought to be the photographer of phenomena; his observations ought to represent nature exactly."[31] In America, Frank Norris reiterated Zola's injunction almost as if he had been fiercely cathecized: "No one could be a writer until he could regard life and people . . . from the objective point of view—until he could remain detached, outside, maintain the unswerving attitude of the observer."[32] And Stephen Crane and Harold Frederic wrote appreciations of each other's work in similar terms of the truth or accuracy of recording: "Like the camera which exposed the romantic distortions of generations of battle painters, Crane's 'photographic revelations' suddenly illuminated the authentic face of war," Frederic wrote of Crane, and Crane reciprocated by praising his friend's craftsmanship and comparing his mind to a "sensitive plate exposed to the sunlight of '61–'65."[33]

The camera provided them with a language of neutrality. It was as if each of them had read James F. Ryder's adventures as one of the first roving photographers and put himself in his place, substituting raw, unbiased consciousness for Ryder's mechanical recording device: "[The camera records] truth itself. What he told me was as gospel. No misrepresentations, no deceits, no equivocations. He saw the world without prejudices; he looked upon humanity with an eye single for justice. What he saw was faithfully reported, exact, and without blemish."[34]

Like Ryder, they wanted to report faithfully, to write exactly and "without blemish." They were concerned with securing external likenesses, fascinated by the permutations of the surface of life, as if literature no longer had reason to render depth, interiority, or the relation of seen to unseen existence. But we should remember that the need to be objective is, at base, an attempted solution to the problem of approach, a problem made more urgent to them by the great, emerging underclass of workers whose experiences increasingly claimed the attention of their peers. "How can one come upon poverty, pain, or deprivation?" is a question that often precedes or underlies the realists' preoccupation with neutrality of stance. To them, it was a position that seemed fair and right-minded; it seemed to negotiate the boundaries of class in the cleanest possible terms; it was precedented in science and in the nascent social sciences. If to us such disinterest seems to cloak a more submerged interest in defining and controlling the subjects of investigation, it is nonetheless important to remember that they did not see themselves in such imperialistic terms.

Although Theodore Dreiser is often considered a realist, his "truths" were arrived at differently and they served another function in his life. His career shows the curious progress of a writer whose dilemma did not arise from the need to approach an unknown Other, but from an excessively close identification with the dispossessed subjects of his art. For him, the need to see, to frame experience, was initially a function of deprivation: it distanced him from a group of undistinguished peers and at the same time removed the distance from the world of his desire. "I stared," he said of his youthful destitution in New York, "with the eyes of one who hopes to extract something by mere observation."[35] If Hawthorne had wanted to appropriate the experience of others through vision, as if sight could satisfy hunger, Dreiser wanted to embrace the world of social position and material ease in a similar way. He did not hold the life of the streets in esteem or consider that it had a mythic source of strength because he had, if anything, too much access to that life. For him, observation was powerful in proportion to its ability to lift him out of destitution, as if perspective itself could free him from necessity. If he began his career with the narrative stance of the disenfranchised facing the world of desire, he eventually moved to a position of greater stasis, a position in which observer and

subject met on grounds of increasing parity. Where observation initially served to exclude him from his own field of vision, it came eventually to be a form of self-reflection, a means of including himself in
address to his own situation as if he had become the figure of John
Berger's imagination—a man of action who could also look at himself
"like the old crow looks at us!"

Dreiser's struggle with the meaning of the objective view, the
unpartisan stance, was not a solitary one. As John Dos Passos, James
Agee, and John Steinbeck were drawn into the great political upheavals
of their times—the Spanish Civil War, the Depression of the 1930s, the
farm labor movement in California—they stood before and in the midst
of experiences which fascinated and appalled them. What was the artist's role in conflict, change, and the restitution of order? Was he or she
the perennial outsider, the "parasite on the drama of life," someone like
Ted Roethke in his elegy for a dead child "with no rights in this matter"? Their dilemmas show the expanding political relevance of the
problems of surveillance that had seemed to Hawthorne so private, so
shamefully personal.

Dos Passos's dramatization of his dilemma in "The Camera Eye"
sections of USA lets us understand the terrible cost of witnessing and
being witnessed. "From trying to pry into their lives as they pass me
on the street—I have reached a point where I feel more like a disembodied spirit than a warm-fleshed human," he wrote to Dudley Poore
in 1916.[36] Later, musing on the other side of this process of acquisition
and transformation, he asked, "Can it be that the Arabs are right, and
the dour pueblo-dwellers of our own Southwest, when they say that the
camera takes away something that can never be recovered, skins some
private value off the soul?"[37] Who benefited from this dynamic? Who
was depleted or "skinned" by it? In what ways did the artist reiterate
patterns of domination and subversion that characterized the political
struggles from which he or she claimed to stand aloof? Agee saw only
the damage of investigation. Steinbeck exempted himself from the
political process but unwittingly replicated the very structures of supervisory authority that he criticized as a tactic of the Growers' Association in repressing migrant labor. All of them circled around the issues
evoked by their own watchfulness; and in understanding its photographic nature, in exploring the machinery of their own vision, they
revivified the meaning of "point of view" as a category in the description of narrative. They historicized it, gave it new resonance, revealed
the political ramifications of their own chosen forms of author-ity.

That vision could have an ultimate form of authority was pointedly
revealed to them in the confrontations engendered by World War II.

With life at risk in the very circumstances that demanded recording, neither photographers nor writers could avoid understanding the predicament of their subjects. As observers, they either shared the threat of death at the hands of a common enemy, or else their very safety forced them to confront their unarguable position as outsiders to the action they sought to represent. In either case, their status was laid bare by extremity.

Margaret Bourke-White was allowed to photograph first from the Fifth Army's surveillance aircraft and later from their bomber planes. Her stance insured that she saw the war from the perspective of those who initiated destruction, and she allowed the remoteness of these views to shield her from the terror of what was contained in them. Above all else, she was overwhelmed by the pageantry before her:

> As we headed toward the front I was impressed with how regular the pattern of war, seemingly so chaotic from the ground, appears from the air. The tracks of pattern bombing at an airfield were as regular as though drawn with ruler and compass.[38]

Robert Capa, who chose to share the experiences of the common GI, photographed from the ground in the midst of combat. For him, war had none of the glittering aesthetic quality that Bourke-White observed, and, in fact, Bourke-White's perspective would have been anathema to him. Destruction, as he viewed it, had nothing lyric about it: "This was the first time I had followed an attack from beginning to end, and I managed to get some good pictures and showed how dreary and unspectacular fighting actually is."[39] His own brilliance lay not only in the "dreary" pictures that let the world see the suffering of common people caught in the machinery of a history they could neither understand nor control, but also in helping John Steinbeck to understand that photography was implicated in history and not a neutral recorder of it. Several years after the war when he watched the nervousness of Soviet citizens with Capa, Steinbeck realized that "in the mind of most people today the camera is the forerunner of destruction."

> The camera is one of the most frightening of modern weapons, particularly to people who have been in warfare, who have been bombed and shelled, for at the back of a bombing run is invariably a photograph. In back of ruined towns, and cities, and factories, there is aerial mapping or spy mapping, usually with a camera.[40]

As Norman Mailer, who ends this study, recognized from his own wartime experiences as an aerial photographic interpreter, his own narrative choices were defined by these extremes. He could write from the aerial or command position where strategies were clear and the individual consequences of them were obscured by distance, or he could bridge that distance through shared experience. To join the armed forces, to risk his life, was for him, as for Capa, a political decision as well as a precondition of his art.

If, as Jay Cantor suggests, we constantly create values by our activities, if "[a]rt and politics are of the same metaphysical substance,"[41] as I believe they are, then the lessons of these writers and photographers are invaluable to us. For they show the conduct of art as a process of life, the making of literature and photography as a constant transposition of values enacted in the world. To think about frames in this context is not to think about museum walls but to remember a negotiation or approach to the subject, which joins history, personal history, and aesthetics. It is to understand that works of art can be blessed or cursed in the same way that their creators have been blessed or cursed by the world of human endeavor.

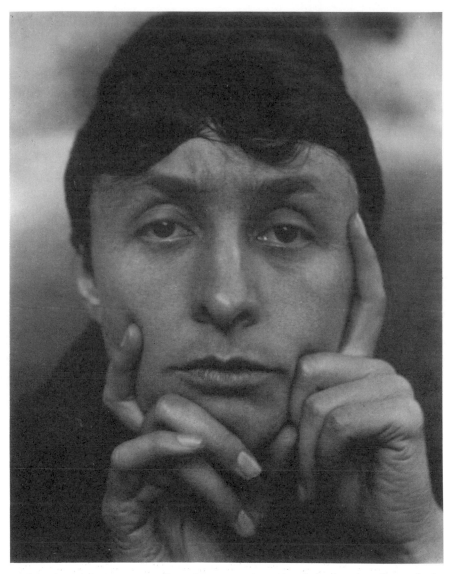

Alfred Stieglitz. Georgia O'Keeffe: A Portrait—Head *(National Gallery of Art, Washington, D.C., The Alfred Stieglitz Collection)*

Don McCullin. Nine-Year-Old Cambodian Boy, Victim of Napalm, 1976. From *Hearts of Darkness* (New York: Knopf, 1981).

J. J. E. Mayall. Nathaniel Hawthorne. London, 1860.

1

Nathaniel Hawthorne
and Daguerreotypy

Disinterested Vision

Thus excluded from everybody's confidence ... I
had often amused myself with watching.
—Nathaniel Hawthorne

I

SOON after Nathaniel Hawthorne returned to America in 1860,
he responded to James T. Fields's request that he contribute an
article to the *Atlantic Monthly*. From Concord he wrote, "Perhaps
I might find some sketch of rural or town scenery that would do. There
is a long account of a visit to places connected with the memory of
Burns—not in the least thoughtful or imaginative, but of the photographic kind."[1]

He had been in England between 1853 and 1857, in Italy for the
next three years, and then in London long enough to secure British
copyright for his latest romance, *Transformation*. He remembered these
years as the most active and in a sense the most strained period of his
life, for his duties as American Consul in Liverpool had given him his
first substantial public or official identity. Accustomed to family life in
small New England towns and to minor commercial posts, he had been
called into service in an unfamiliar place; and though he had frequently
felt estranged from his own life circumstances in America, England,

with its round of unwonted duties, had elicited a pronounced sense of exile.

To Fields, of course, these wanderings abroad held no similar anxiety. He and his fellow editor, William D. Ticknor, had known for years that Hawthorne's experiences were well documented, for he had written to them that his notebooks "could easily make up a couple of nice volumes."[2] Fields presumably was asking for something very simple: travel journalism in a conventional mode and from a man reputed to be a thoughtful and imaginative observer. Yet Hawthorne replied that he proposed to write essays "not in the least thoughtful or imaginative."

By saying this, Hawthorne probably wanted to tell his publishers what to expect from him: he usually wrote fiction, and he presumably wanted to discriminate between one kind of writing and another. But what was it that brought him to make that distinction in terms of photography? As he proposed to write non-fiction, did the photograph (or daguerreotype) represent truth from which imagination deviated in some measurable way? We know that he feared his notebooks' candor and that they were, according to his own assessment, "too spicy."[3] They would "unluckily . . . be much too good and true to bear publication. It would bring a terrible hornet's nest about my ears."[4] Did he think his pieces would be superficial and want to warn Fields that he would send him a dreary record of surface details? In this regard we know that he did not particularly like the collection of pieces that Ticknor and Fields eventually published as Our Old Home (1863), judging that it was "not a good nor a weighty book . . . deserv[ing] any great amount either of praise or censure."[5] Did he want to call attention to the objectivity of the sketches, to their lack of sensibility and private discrimination? Could he have referred to their origins, either in the 300,000-word notebooks or in life, and emphasized their fidelity to a prior source? In comparing the finished essays to those notebooks, we can discover a variety of relationships—literal transcription, polish and selection, embellishments from a later vantage point in time, omissions for the sake of discretion. Was the word "photographic" used as a metaphor rather than as a simile? Was he himself the camera turning a scrupulous eye on the life spread out before him, a mechanical and indiscriminate transcriber of an alien scene?

A simple distinction can give rise to a wealth of speculations but in this instance it is probably safe to guess that Hawthorne used the word "photographic" without much premeditation and in a common-sense way to mean "true to life." By 1860, photography had become a common feature of New England life, and it was certainly common to

Hawthorne, who had ritually visited daguerreotype studios for portraits since 1841.[6] Yet photography has an uncanny fitness in association with this man, or for that matter, in association with any writer who has meditated on the nature of detachment and on the problems of relationship established through vision. Whether Hawthorne thought of himself as a camera, or would have thought there to be virtue in such a posture, is not important to determine in a literal sense. For it is clear that his own reticence led him to become a writer whose primary resource, like the photographer's, was sight, and whose art was predicated on material appropriated from those distant visions.

Where other artists might have relied on their own experience as the primary material for imaginative transformation, Hawthorne moaned about the dearth of incident in his life and complained that he had to weave his tales out of thin air, to substitute speculation for substantial, informing emotion. Usually the motives and consequences of such marginal living are personal; but for a writer like Hawthorne, the consequences of estrangement were aesthetic as well. For if deprivation created isolation and pain, it also necessitated a kind of visual greediness in the pursuit of art, a need to know predicated on the desire to write, a voyeurism that unhappily seemed never to be definitive.

In his concern about vision, Hawthorne was hardly alone; in fact, he lived in an age that was preoccupied with optics. His friend Emerson articulated the sentiments of many New Englanders when he observed that sight was the most spiritual of all the senses. Although Hawthorne was undoubtedly intrigued by Emerson's conflation of sight and insight, he was equally certain that he did not know what he had got when he finished looking at anything. For such a man, the camera was bound to have special associations, if not as an analog to the form of his narratives, then as a metaphor whose terms contained and expressed the problematic relationship of seer to seen, and the concomitant anxiety of uninvited scrutiny.

Hawthorne's remark to Fields, then, was not inconsequential, and it assumes added meaning when one remembers that Hawthorne began his writing career as a self-conscious observer, that looking-on was not a habit he had picked up to deal with unusual circumstances in England. Although marriage, children, professional work, and finally his own literary reputation intervened to bring him both the pleasures and costs of involvement with a community, his youthful detachment was a posture he never lost, and which never lost its anxiety for him. In fact, many of his earliest sketches dramatize the nature of his isolation, his subsequent ignorance about people, and the broken and unsatisfactory attempts he made to breach psychological distance by watchfulness. Many readers have considered "Sights from a Steeple" to be one of his thinnest pieces of narrative, but it is also one of the baldest betrayals of Hawthorne's desire to see without being seen, to know without knowledge exacting a cumbersome price. Through it, we can see Hawthorne working out the implications of his own tendencies toward voyeurism.

Let us look at the sketch for a moment. In it, an unnamed narrator climbs up into a church steeple and looks out over a town, observing the meeting of lovers, the commerce of the wharves, the marching of soldiers, the play of children, the line of a funeral procession. In a certain sense, the sketch's economy bears comparison to Emily Dickinson's poem, "Because I could not stop for death," where the narrator's journey toward the grave takes her past all of life's remarkable activities. In Hawthorne's vignette, the narrator's attention is arrested by one particular incident—the morning walk of a young man, his meeting with two girls, his obvious attraction to one, and his disappointment when the girls' father interrupts their good times to take them in out of the impending rain. Hawthorne describes the young man as he walks up the street: "He saunters slowly forward, slapping his left hand with his folded gloves, bending his eyes upon the pavement, and sometimes raising them to throw a glance before him. Certainly he has a pensive air . . ." and speculates about the meaning of these details: "Is he in doubt, or in debt? Is he, if the question be allowable, in love? Does he strive to be melancholy and gentlemanlike?—Or is he merely overcome by the heat?"[7]

What is remarkable about the sequence is Hawthorne's keen sense of what can and cannot be deduced from looking-on: he can see and describe the young man's appearance; but since no answers are possible to his questions about motivation, the narrator turns his eyes in other directions, returning to the same spot only when the man meets the young women and walks on with them. From the scene he constructs a story of sorts—that the young man loves one of the sisters—but it is an unsatisfactory concoction not only because it is unverifiable, but also because his guess is in fact a projection of his own prior attraction to the woman, and as such, it is a covert admission of what he is losing by watching her instead of wooing. Though he describes himself as "all-heeding and unheeded," he is in fact more involved in the scene than he admits.

This edge of personal involvement lends curiosity to his wish that the "chimneys could speak . . . and betray . . . the secrets of all who . . . have assembled within" and to his later remark that *the most desirable mode of existence might be that of a spiritualized Paul Pry, hovering invisible round men and women,* witnessing their deeds, searching their hearts, borrowing brightness from their felicity, and shade from their sorrow, and retaining no emotion peculiar to himself [my italics].

But none of these things are possible. And if I would know the interior of brick walls, or the mystery of human bosoms, I can but guess.[8]

These are the comments of a man for whom vicarious experience is all there is of experience, a man whose emptiness demands compensation in the external world. It could also be claimed that they express the wish to expropriate emotions without involvement and without the guilt that usually accompanies more overt or calculated manipulation.

While the narrator is busy gazing, he remains poised, but the sketch is brought abruptly to an end by a thunderstorm that sends the villagers into those very brick houses that sight cannot penetrate. When he surveys the scene for the last time, the observer remarks that it is a town "whose obscured and desolate streets might beseem a city of the dead," and turning a single moment to the sky, he adds that it is "now gloomy as an author's prospects."[9] Without vision the narrator is thrown back on his own isolation; without vision he has nothing to write about. The storm has robbed him of his story, and so it ends. This, put simply, is the central aesthetic problem of the sketch. It is an unsettling vignette both about personal hunger and about the preconditions of art as Hawthorne understood them.

"Sights from a Steeple" was first published in book form in 1837. It does not render experience in terms of photography for the good reason that cameras, as we know them, had not yet been invented. But in it we can see Hawthorne searching for a way to represent psychological distance that does not preclude connection to the world. Here he chooses to indicate that relationship with a tower; in other sketches of *Twice Told Tales,* he makes use of windows, or in one case, he wishes himself transformed into a toll-gatherer who sees life parade past his booth. It is the seed of an obsession. Hawthorne may have later taken his characters out of the tower and let them walk down streets into those problematic brick houses or through pilgrim forests, but the sensibility that observed and that needed to be fed by observation remained and was recurrently dramatized in his fiction.

For our purposes, the question raised by this text concerns Hawthorne's eventual association of such aesthetic detachment with cameras, a question whose answer lies as much in the history of photography as it does in Hawthorne's particular sensibility; for the associations that came eventually to mind were to some extent the associations of an entire era.

II

In March 1840, François Gouraud arrived in Boston. A pupil of Daguerre's, he came to sell daguerreotype pictures, to market a technique and its apparatus, and to convince a lively but conservative city that "photogeny" deserved its attention. He set up both private and public exhibitions and invited the notable men of the town to an exclusive showing:

> Having brought with me a choice collection of the most beautiful specimens of my Daguerreotype drawings, I have thought it my duty, before exhibiting them to the public, to give some of the most eminent men and distinguished artists of this city, the first view of perhaps the most interesting object that has ever been offered to the examination of a man of taste—and therefore, if agreeable to you, I shall be happy to have the honor or receiving you ... at Tremont House, in the ladies' Drawing-Room.[10]

Judging from his announcement, he was genteel, debonair, and rather like P. T. Barnum with a good side-show on his hands. While Gouraud was teaching and lecturing in Boston, he competed with the lecture circuit of the Society for the Diffusion of Useful Knowledge and with the New York and Boston Diaramic Association which advertised "mysterious and mechanical moving diaramics of animated nature."[11] That photography was originally situated in print between these two alternatives of didacticism and entertainment was not as haphazard as it first may seem, for the impact of so new an invention could not immediately be measured. Was it a gadget, to be associated with optical illusions, or was it a significant cultural achievement? Did it deserve the attention of the elite or the masses? Herman Melville was to dismiss the daguerreotype as another pasteboard mask; Emerson was to laud it as one of the five miracles of his lifetime. Throughout Boston responses to Daguerre's invention would remain similarly varied.

On one extreme were those like Albert Southworth and Josiah Hawes, who saw personal opportunity in Gouraud's demonstrations. These men were both young, from relatively unestablished families, and in the position of starting careers at an economically depressed time. Gouraud offered them a profession that required little capital investment and the promise of a large, relatively uncompetitive market. The first entry in the account ledger of their firm was in June 1841.[12] They had mastered an uncharted trade and set up offices in Tremont Row in little over a year. Elegant surroundings, personal service, and expensive products ensured an elite clientele, and the faces that are mirrored in the surviving plates from their studio tell us that they persisted in style: Oliver Wendell Holmes, Ralph Waldo Emerson, Henry Wadsworth Longfellow, Nathaniel Hawthorne, and Daniel Webster all sat for their pictures here, and often allowed their images to advertise the expertise of the photographers—a practice Hawthorne was to tease gently in *The House of the Seven Gables* by having his fictional daguerreotypist solicit a publicity photo from Uncle Venner, the town's half-wit.

From the researches of Beaumont Newhall[13] and Robert Taft[14] we know that Southworth and Hawes were quickly joined by other salons and by itinerant tradesmen who offered to the general public less expensive but sometimes equally attractive portraits. As several men of the times were to notice—Emerson and Holmes among them—photography was always a potentially democratic art, giving to middle class and poor people the permanent images that had previously been an elite prerogative. At $2 per sitting, almost anyone could afford a daguerreotype. Between 1840 and 1860, one hundred eight daguerreotypists were listed in the Boston City Directory, and though this is probably an unstable number—the turnover in the trade was high and failure was common—it indicates to some extent how quickly photography became a feature of New England life.

Whatever the fortunes of any individual photographer, the medium did attract general interest. Edward Everett Hale, who went to one of Gouraud's first lectures while he was still an undergraduate at Harvard, became an early photographic experimenter, puzzling over chemicals, timing, and lighting procedures as much as Southworth and Hawes, although for him photography remained an avocation rather than a profession.[15] In old age, he remembered that Francis C. Gray had arranged for Gouraud to give demonstrations at the Massachusetts Historical Society and that the Dixwell family had contracted for private family lessons which he was invited to attend.[16] His involvement in photography was incidental to a life preoccupied with the ministry and issues of social reform, and though he was unique among men of his class for practicing the craft, for actually knowing its technical intricacies, it remained incidental to his habits of mind as well.

Others who merely sat for their portraits—playing the passive role—often had more interesting thoughts about the subject. These opinions, incidental though they are, are particularly important because they apparently do not stem directly from Gouraud. The public might have simply parroted back the attitudes of a learned man on the lecture circuit, but since Gouraud's manual of instruction was not a treatise on the nature of photography but only a step-by-step account of a technical procedure, this did not seem to have happened. Paperbound in plain green wrappers, printed by Dutton and Wentworth in 1840 to accompany his lectures, the booklet divided the daguerreotype process into five separate parts. In retrospect it seems clear that people's imaginations were stirred as much by the procedure of taking pictures as by the nature of the finished products. These were the operations:

1. cleaning and polishing the plate;
2. applying a sensitive coating to the plate;
3. putting the plate into the camera, exposing it "to the action of light, for the purpose of receiving the image of nature";
4. bringing out the invisible image with chemicals; and
5. removing the sensitive coating to prevent further solarization.[17]

Each of these steps involved the intricate manipulation of chemicals, buffers, heat, and light, along with mechanical dexterity and an instinct for the right moment and posture; but what is most interesting about the early attitudes toward photography—among photographers and laymen alike—is their almost uniform failure to regard the photographer as an active participant in, and shaper of, procedures and events. Instead commentators stressed the necessary mechanical connection between the visual image and what was in front of the camera. No matter how sophisticated the chemical analyses or how detailed the accounts of sitting arrangements that appeared in manuals, photographers tended to see the form as autotelic and to ignore their role as manipulators and originators. Gouraud thought he was showing how "nature impresses an image of herself,"[18] and most of his followers reiterated this language of human passivity, of automatism. At best the photographer enabled the sun to draw images of the world; he was a catalyst of light, the conduit of a phenomenon that was thought to exist independent of human agency.

Nathaniel P. Willis, writing for the New York magazine *The Corsair* in April 1839, announced, "[A]ll nature shall paint herself."[19] Even Fox Talbot, who invented the photographic procedure that left its image on paper rather than on metal, entitled his manual *The Pencil of Nature* and explained that it was a "process by which natural objects may be made to delineate themselves."[20] Samuel F. Morse, Daguerre's

American friend and patron, expanded these sentiments by calling daguerreotypes "*fac-simile* sketches of nature." They were "painted by Nature's self with a minuteness of detail which the pencil of light in her hands alone can trace. . . . [T]hey cannot be called copies of nature, but portions of nature herself."[21]

No other invention in a time of splendid inventions (which included the railroad, the telegraph, and the steamboat) was so quickly pulled away from the particular genius of the inventor and the ingenuity of his followers. The camera and the daguerreotype were quickly, almost instinctively, fit into a system of thought that recognized truth in a certain style of nonsubjective vision. Since nature recorded her own appearance, since no individual sensibilities interfered with its rendering, photographs were thought to be true.

One could see in these attitudes an ironically twisted return to the classical concept of the artist, the view commonly held before Dante, Petrarch, and the writers of bourgeois Italian culture, where only the manual skill of the artist was appreciated since his creative power was credited to God. The photographer might be the dextrous handler of chemicals and lenses, but nature was the real artist working through him. Or one could place these attitudes in another more immediate context (one that is in an odd way not completely antithetical to the above) and understand them as responses to contemporary science, as an intuitive alignment of photography with the objective postures beginning to be required of all natural observers. Both views discouraged the creative role of the artist or scientist. It should be remembered that Daguerre was one of many experimenters on the nature of optics who commonly assumed that valid knowledge must not be mixed with subjectivity. The photograph was invented at a time when "fact" was conceived of as sensory evidence at the same time that perceiving subjects were required to be eliminated—a contradiction of early positivism that has only received retrospective attention.[22] In the mid-nineteenth century the photograph seemed to be just that—a "fact" acquired without a perceiving subject—and this view persisted alongside of other attitudes that seem contradictory to us. For example, Anson Clark, writing in 1841, observed:

> The value of a portrait depends upon its accuracy, and when taken by this process it must be accurate from necessity, for it is produced by the unerring operation of physical laws—*human judgment and skill have no connection with the perfection of the picture,* any more than with that formed upon the retina of the eye, and the likeness produced will be the exact image of the object [my italics].[23]

But it is worth noting that while Clark insisted on the inevitable accuracy of the image, he knew full well that some of the initial problems of studio photography were ones of creating plausible illusions. As early as 23 October 1839, the New York *Evening Post* listed as improvements in daguerreotypy a series of faking devices: one could paint the subject's face white, powder his hair, and hold his neck in a vice to get better resolution and reflection; and though these extraordinary measures were abandoned when more sensitive plates were developed, photographic manuals throughout the decade retained numerous suggestions for posing subjects—arranging draperies, choosing backdrops, adjusting posture. In short, they habitually prescribed the kind of human interference and acts of judgment that belied the theoretical descriptions of their work and account for some of the characteristics of daguerreotypy that now seem most representative: staged appearances, frozen inert bodies, rigid expressions, and premeditated compositions of light and shadow.

It is initially difficult to know what to make of these contradictions beyond noting that they were pervasively held. Although the photographer, Marcus Root, thought that the human face was an index of character, the body a reliable sign of the soul, he freely admitted that few daguerreotypes displayed the "intelligent, spirited, noble cast of countenance which we instinctively ascribe to the originals."[24] The comments of many people who sat for their portraits bear this out. Emily Dickinson, John Quincy Adams, and Harriet Beecher Stowe almost invariably hated their photographic images, and Stowe in particular thought them "useful like the Irishman's guideboard which showed 'where the road did not go.' "[25] Clearly the best arrangements of the time could not dispel every dissatisfaction about the nature of photographic representation.

There are several ways to untangle these loose juxtapositions of contrary thought. In the case of studio tampering, photographers apparently maintained a distinction between the process of recording and the nature of the object viewed through the lens, reserving for the one an absolute verity and for the other a range of acceptable manipulations. Once the shutter was open, a camera could not choose what to record, and in this mechanical sense gave true evidence; but what was placed before it could be selected for a variety of reasons other than candor. This distinction is a simple one, and it is less interesting than the problem that stems from the dissatisfaction of subjects who sat before the camera in all honesty, presenting straight faces to the world and hoping for a "good" likeness. Their disappointments need not express foiled vanity—though that was almost certainly part of it—but a less easily expressible feeling that their pictures had not "captured" or represented a sense they had of themselves that existed prior to the sitting. Though they could not fault the daguerreotype image in particular—the recorded features were undoubtedly their own—they nonetheless retained a sense that these externals were "true" in very limited and particular ways. They perceived a discrepancy between what they saw represented in a plate and what they felt to be unexpressed in their own character.

Broadly speaking, those who rejected the evidence of photography, or thought it to be partial truth, did so because they felt that their images obscured rather than revealed "the inner man." Richard Rudisill[26] has suggested that most Americans of the daguerreotype era interpreted photography in light of their own transcendental concerns, and automatically "read" daguerreotypes symbolically. He has tried to make the case that daguerreotypes were thought to be "natural facts" and as such were the inevitable corollary of spiritual facts, examples, in short, of the situation Emerson describes in "Nature": "Every appearance in nature corresponds to some state of mind, and that state of mind can only be described by presenting that natural appearance as its picture."[27] But this correlation assumes that the nineteenth century inevitably conflated the daguerreotype image with the natural objects that it purported to represent;[28] and it presupposes a uniformity and certainty of opinion that does not seem to have existed.

Certainly there were those who thought the truth of the camera to be complete in a spiritual as well as a material sense. Marcus Root said quite plainly that the human face as represented in the daguerreotype was the mirror of the soul. James Ryder also reified his camera, thinking of it as a machine gifted with the discernment of character: "The box was the body, the lens was the soul with an 'all-seeing' eye, and the gift of carrying the image to the plate."[29] "He could read and prove character in a man's face at sight. To his eye, a rogue was a rogue; the honest man, when found, was recognized and properly estimated."[30]

For all those who retained undivided faith in the fullness of the camera's representation, there were many others, however, who worried that the image recorded information that was very superficial. John Fitzgibbon, the inventor of the "Arkansaw Traveler," had his character say, "Picters can't show the innard man ... that's the part I vally,"[31] and Oliver Wendell Holmes retained a similar light vein when he spoke of the photographer as a great white hunter collecting the skins of objects.[32] But their humor need not obscure the problematic nature

of the issue. Thoreau, in one of the few reflections about photography
in his journals (February 1841), is clearly concerned with the relation
of exterior to interior:

> It is easy to repeat but hard to originate. Nature is readily made to
> repeat herself in a thousand forms, and in the daguerreotype her own
> light is amanuensis, and the picture too has more than a surface sig-
> nificance,—a depth equal to the prospect,—so that the microscope
> may be applied to the one as the spy glass to the other. Thus *we may
> easily multiply the forms of the outward; but to give the within outwardness is
> not easy* [my italics].[33]

Emerson said things initially that suggest he thought of photographs as
representations that joined inner and outer symbolically: "The daguerreotype is good for its authenticity. No man quarrels with his shadow,
nor will he with his miniature when the sun was the painter. Here is
no interference, and the distortions are not the blunders of an artist,
but only those of motion, imperfect light, and the like."[34] But when he
came to sit for his portrait and experienced the discomforts of enforced
rigidity, he too began to question the depth and acuteness of what was
recorded: "[U]nhappily, the total expression escaped from the face and
[you found] the portrait of a mask instead of a man. Could you not by
grasping it very tight hold the stream of a river, or of a small brook,
and prevent it from flowing?"[35] Melville was the most skeptical of his
peers, resolutely dismissing the daguerreotype as he did most of the
material world of flimsy appearance.

Each of these men brought to photography a predisposition to read
the natural world symbolically; each of them initially thought of the
daguerreotype as he had been taught to think of it—as the representation of an inner state through signs given in external experience—and
each of them apparently came to feel uncomfortable with that simple
equation. Their comments about the discrepancies between outer and
inner, between the frozen representation and its lively referent in the
physical world, suggest that photography's success as a sign of spiritual
life was not uncomplicated.

Of course they have left only fragmentary responses, but the questions raised in them tell us that not all of the common explanations of
daguerreotypy rested easily. To observe this is not to cast doubt on the
earnestness or intelligence of photography's first advocates. Morse,
Root, and other experimenters explained what they thought to be the
case—that the sun was principal agent in the process, the chief "artist,"
and that it invariably told the truth. By their peculiar understanding
of the nature of a process, they immediately gave a special scientific or

at least intersubjectively valid status to the images recorded. Given the context of scientific values they participated in, it was natural for them to want a procedure that circumvented human agency, that recorded the physical universe without the interference of fallible intelligence. But between these declarations of truth and the hesitant responses of individuals to their own "true" portraits, one can see a beginning matrix of problems surrounding the role of the photographer, his methods of observation, and the nature of the representation obtained by a supposedly impersonal process.

What kind of truth were people confronted with in these small metal plates, which on the one hand delineated features with incomparable finesse, and on the other disappeared from sight when held at the wrong angle?

III

These complicated issues provide the context for Hawthorne's responses to photography. He, along with Emerson, Melville, and other of his New England peers wondered about the nature of Daguerre's small metal plates. But despite the evidence of Hawthorne's eventual interest, it is difficult to discover how he first came to know about photography. If he was invited to Gouraud's first exclusive lecture along with Longfellow, Edward Everett, and Boston's other elite, the letter of invitation no longer exists to prove it. To hazard a guess, it seems unlikely that he was included, for in 1840 he was measuring commodities in the Boston Custom House and writing morose letters to Sophia Peabody about his obscurity. Since he was in the city, and in fact living around the corner from Tremont Row, it is conceivable that he attended the public showing at the Horticultural Hall or the general lectures which Gouraud gave at the Masonic Temple; but again, this must remain speculation.

A more likely connection to the lore of photography might have been established through Oliver Wendell Holmes, whose essays on photography in the *Atlantic Monthly* had influenced general public attitudes, and who became as close a friend as Hawthorne tended to have.[36] One can guess they talked about photography together—they had known each other since 1843 and had met often enough in the Berkshires—but it is not possible to document those conversations. It is clear that Holmes's essays could not themselves have served as the first source of Hawthorne's knowledge or interest, for Hawthorne had already created his protagonist Holgrave, the daguerreotypist, in 1849—long before those *Atlantic* essays were published in 1859, 1862, and 1863. The other keynotes of their friendship also occurred later in life when they shared Ticknor and Fields as a publisher, attended meetings of the Saturday Club together, and watched their two sons become friends at Harvard College. Holmes wrote one of Hawthorne's obituaries and later wrote another appreciation of him to accompany a story published posthumously in the *Atlantic*. He had long been one of Hawthorne's admirers, and was one of the people whose opinion Hawthorne had most valued when *The House of the Seven Gables* was published in 1851. To the gift of a first edition copy of it, Holmes had replied enthusiastically, "pointing out a hundred touches, transcriptions of nature, of character, of sentiment, true as the daguerreotype," though this evaluation did not prevent him from classifying Hawthorne as the country's foremost romancer and seeing that "the solid reality and homely truthfulness of the actual and present part of the story are blended with . . . weird and ghostly shadows."[37] By 1851, photography was clearly a subject both men were familiar with, though the direction of influence is difficult to determine.

However Hawthorne may have first learned about photography, the numerous daguerreotypes and photographs that still exist tell us that he shared his peers' enthusiasm for preserving images of himself and his family. In 1874, years after Hawthorne's death, Thomas Higginson called on Hawthorne's daughter, Una, who was then engaged to marry his nephew, and recalled that the plainness of Hawthorne's house in Concord was punctuated only by photographs and little ornaments on the mantelpiece.[38] We know that Hawthorne occasionally had his portrait painted, but on the whole, he was one of numerous Americans whose budget stretched infrequently to more substantial original art. During his lifetime, Hawthorne was not chosen to be among Mathew Brady's "Gallery of Illustrious Americans" (1850), but he did go privately to Brady's studio in New York, as well as to the studios of Southworth and Hawes (a letter from Sophia Peabody to Hawthorne on 30 May 1841, refers to a miniature daguerreotype that lets us know that he was one of Southworth's first customers); Silsbee, Case and Company; Warren Portraits in Boston; and to the Mayall Studio in London. John A. Whipple photographed him, as did W. H. Getchell and various itinerent cameramen who "seized" him on the streets. Una, Julian, and Rose, his three children, were photographed from childhood on.

Had Hawthorne merely sat for his daguerreotype portrait, he would have no unique claim to our attention in relation to photographic history, for such portraits were a craze in most American cities during the 1840s and 50s. As an article on "The Daguerreotypist" in *Godey's Lady's Book* of 1849 (the year Hawthorne began work on *The House of the Seven Gables*) observed, "If our children and children's children to the third and fourth generation are not in possession of portraits of their ancestors, it will be no fault of the Daguerreotypists of the present day."[39] Holmes reinforced this view by quipping that daguerreotype calling cards were like greenbacks—the latest social currency.[40] But two circumstances, one of them incidental and another of more serious consequences, lend Hawthorne's experiences special weight: he was one of the first American authors whose work was illustrated with photographs, and he was also one of the first of our authors to weave daguerreotypy into his fiction.

The first of these circumstances is partly clouded in mystery, and we will have to remain satisfied with spotty information about the publishing history of the Leipzig edition of *Transformation (The Marble Faun)*. Instead of a full correspondence, we are left with one letter, with artifacts, and with what can be deduced from examining them. We do know that sometime during his stay in England in 1860, Hawthorne entered into a curious arrangement with Bernhard Tauchnitz to publish an English language edition of his work for German readers and for American visitors to the continent. Four years later, in 1864, Tauchnitz wrote to Hawthorne, sending him royalties from *The Scarlet Letter,* and reminding him that he and his London publishers, Smith, Elder and Company, had also consented to the German publication of *Transformation*.[41] It is impossible to reconstruct Hawthorne's role in planning the edition, but the book appeared in two volumes in 1860, illustrated with 57, 59, or sometimes up to 102 photographs pasted individually in place. Since photolithography had not been invented, they are all original photographs (usually of Roman art, architecture, and "typical" Latin character types); and consequently the photographs for each volume are completely unique. No set duplicates another set. One could guess that Hawthorne had no hand in this highly laborious project, that it was the work of an energetic and innovative publisher. We know, too, that the illustrations were, at least in some cases, optional, that the reader could buy sets of one hundred photographs in Italy for twenty-five lire.[42] But the frontispiece of at least some of the volumes is a photograph of Hawthorne (J. J. E. Mayall, 1860, London: the so-called "Bright-Motley" pose)—a circumstance that suggests the author's collusion in the enterprise. How else, in an age that had no publicity agencies, did Tauchnitz get Hawthorne's picture?[43]

Although this edition is an interesting facet of publishing history, it is not possible to claim too much from it. Its existence may bear witness to Hawthorne's extraordinary interest in photography, or it may simply represent his compliance with someone else's idea of pictorial novelty. One can speculate that the images amused him, for though they were truthful representations of something or somebody, they could not, of course, have been true records of his characters. At most one could say that they were visual clues (and in each volume a different visual clue) given in the real world, which guided the reader in imagining the author's fantasy. These are, of course, speculations, but they are not out of keeping with the thoughts about photography Hawthorne did record in *The House of the Seven Gables* where he addressed both the aesthetic and ethical issues photography raised for him as a writer of fiction.

In the preface to the book, Hawthorne looked at the question of formal resemblance between photography and literature. Where photography, as he understood it, faithfully transcribed scenes, rendering proportions of light and shadow with necessary, because mechanical, accuracy, he himself had no interest in exactitude. Instead he described himself as a stage manager manipulating his "atmospherical medium as to bring out or mellow the lights and deepen and enrich the shadows of the picture."[44] He would be inventive, would recognize no obligation to render the lights and shadows of the actual world, would, in fact, render that world with only sparse detail. Henry James thought this paucity of detail to be a mark of the impoverished culture in which his predecessor lived and wrote, but Hawthorne would have disagreed with James: he responded to photography by rejecting it as a formal model for his work.

Yet despite this disavowal, *The House of the Seven Gables* is largely about vision, about the interpretation of visual signs, and about values rendered in terms of light and shadow. And though Hawthorne disclaimed the importance of Holgrave's career as a daguerreotypist ("His present phase, as a daguerreotypist, was of no more importance in his own view, nor likely to be more permanent, than any of the preceding [careers]" [145]), how could he not have recognized that the mid-nineteenth-century language which described the photographic process in terms of sun and shadow, light and dark, and unerring truth, fit with a beautiful logic into his own system of private concerns about art and the ethics of artistic creation? If we look at the text, we can see that Hawthorne placed his daguerreotypist's strange love story in a well-established constellation of private preoccupations, which, when sorted out, can tell us how he distinguished daguerreotypy from other arts, how he responded to contemporary theories of photographic interpretation, and how he related photography to his own situation as a text-

maker. Through Holgrave's profession, Hawthorne recorded his understanding of the photograph as an artifact and his anxieties about photography as a process of human interaction.

The book is essentially a grim meditation about the effects of ances-
tral sin. It presents a theme that Hawthorne used frequently, yet the
story is distinguished by the optimism of its ending. Where many
Hawthorne tales project a continuing series of dire consequences into
the future, *The House of the Seven Gables* ends with a ritual release, the
breaking of a curse, and an escape from the past. The relief afforded by
this escape can be measured against the gloom that originally characterizes the protagonists, for the story begins in an old house built on
stolen land, passed from generation to generation, where Hepzibah
Pyncheon lives with a cumbersome sense of the past. Her solitary desperation is interrupted only when her young cousin arrives for a visit
and when her brother returns from long years in prison. Wrongly con-
victed of murdering an older relative, Clifford returns as a dim and
broken old man who is not able to see what is of most concern to
himself—the reason for his unjust punishment.

For this clarification Hawthorne relies on other characters, and in
one sense, his novel can be read as a modulation of insights, a calculated
array of characters who attempt, with varying success, to penetrate and
understand the nature of the world they see. If Clifford's fate illustrates
the ramifications of misjudging that world, of punishment given
according to poorly understood evidence, then his salvation is predicated on finding a better sleuth, a more astute interpreter of phenomena—someone, in short, who is able to find the real culprit. In this
matter, Holgrave the daguerreotypist acts as Hawthorne's detective. As
a house boarder, he can see whatever he chooses of Pyncheon family
life, yet his position in the house is extremely ambiguous: ostensibly
he is a visiting daguerreotypist, an outsider with no rights in the household. He is perceived as a stranger, yet Hawthorne reveals to us that
Holgrave has a secret motive for being there—he is the unrecognized
descendant of Matthew Maule, whose land the original Pyncheons had
stolen. Unlike Hawthorne's previous observer in "Sights from a Steeple," Holgrave lives inside impenetrable brick walls, but though vision
is not obstructed, he watches people who are already acutely sensitive
to scrutiny. In starting her cent shop, Hepzibah is "tortured . . . with a
sense of overwhelming shame that strange and unloving eyes should
have the privilege of gazing" at her. "It might have been fancied,"
Hawthorne adds, "that she expected to minister to the wants of the
community unseen, like a disembodied divinity" (140). The old maid
locates her nemesis outside the house, in the curiosity of strangers, but

for Hawthorne the more crucial issue is contained within the house
itself, where Holgrave snoops in secret: "In the artist's deep, thoughtful,
all-observant eyes, there was, now and then, an expression not sinister,
but questionable; as if he had some other interest in the scene than a
stranger, a youthful and unconnected adventurer, might be supposed to
have" (130). In fact, it is Hawthorne's interest in the way knowledge is
acquired that eventually joins Holgrave's profession as a daguerreotypist
with the core issues of recognition and release in the text. For if the
young man sees the life around him, his camera sees as well, recording
in permanent images what others encounter as incidental impressions.

Here we see Hawthorne picking up and using the popular attitudes
of his peers, not only in regard to the daguerreotype image, but with
the character of his young man as well. James Ryder, Hawthorne's
contemporary, remembered his fellow daguerreotypists as a restless, ill-
assorted bunch, a group of hucksters whose products were as salable as
cough medicine and miracle cures. "It was no uncommon thing," he
said, "to find watch repairers, dentists, and other styles of business folk
to carry daguerreotypes on the side."[45] Other contemporary accounts
support this view: "Today you will find the Yankee taking daguerreotypes; tomorrow he has turned painter; the third day he is tending gro-
cery, dealing out candy to the babies for one cent a stick."[46] Holgrave
is, then, a recognizable type:

> Though now but twenty-two years old ... he had already been, first,
> a country schoolmaster; next, a salesman in a country store; and either
> at the same time or afterwards, the political editor of a country news-
> paper. He had subsequently traveled New England and the Middle
> States, as a peddler, in the employment of a Connecticut manufactory
> of cologne water.... His present phase, as a daguerreotypist, was of
> no more importance in his own view, nor likely to be more perma-
> nent, than any of the preceding ones. It had been taken up with the
> careless alacrity of an adventurer (143-44).

If Hawthorne's superficial description of Holgrave's past fits popular
photographic lore, his further characterization does not, for Holgrave
is uniquely troubled by his activities. Where other daguerreotypists
traveled, lived routinely in the homes of private citizens, and set up
temporary studios much as Holgrave does in this text, they tended to
find this life rich in human incident. James Ryder, as a repeated example, thought of his camera as a means of wholesome personal experience
and as a companion in adventure. But Hawthorne's daguerreotypist
remains aloof, and his camera mystifies personal motives.

Although a daguerreotype portrait of Judge Pyncheon eventually clarifies the mystery of Clifford's past, it is not, in Hawthorne's eyes a simple artifact, for it is taken in stealth and is the occasion for guilt. Hawthorne felt this even though he distinguished clearly between the photographer's own vision and that of his camera: where Holgrave's observations might be culpable and fallible, his camera's evidence is not. In making this distinction, Hawthorne is following the trends of his time and attributing to daguerreotype images an absolute fidelity to nature. In a sense, it was a natural inclination for a man who struggled throughout life to understand the nature of visual clues and for whom opaque surfaces presented repeated cause for anxiety. As F. O. Matthiessen tells us, "Hawthorne 'abhored' mystery, he dreaded any 'unintelligible expression' as a clouding veil 'between the soul and the truth it seeks.' "[47] Camera lore provided him with a comforting, if gimmicky, solution to the plot of *The House of the Seven Gables,* where breaking Maule's curse is a function of correctly assessing Judge Pyncheon's character. This is an ability that none of the characters individually possesses since the judge strives to obscure his own maliciousness under a mask of benign respectability. Phoebe stands in awe of his authority, Hepzibah and Clifford fear him so intensely that they cannot make any plausible connection between their fears and an objective situation. Holgrave suspects the judge's role in ruining Clifford's life; he suspects an evil nature clothed in sunshine, but he cannot prove it—and amid all of Hawthorne's customary waffling with alternative explanations, proof is precisely what is needed to make a definitive break with the past and to free Hepzibah, Clifford, and Phoebe from the weight of wrongly inherited guilt. The rescuer must be capable of extraordinary discernment. In fact, he must possess precisely those abilities that James Ryder attributed to his daguerreotype camera: "He could read and prove character in a man's face at sight. To his eye, a rogue was a rogue; the honest man, when found, was recognized and properly estimated."[48]

Holgrave's camera, not surprisingly, has similar powers of observation. "There is a wonderful insight in heaven's broad and simple sunshine. While we give it credit only for depicting the merest surface, it actually brings out the secret character with a truth that no painter would ever venture upon, even if he could detect it" (80). While Phoebe dislikes daguerreotype images for the same reason Emerson did ("They are so hard and stern [and] unamiable"), Holgrave counters with a stereotyped avowal of their truth: "The likenesses may be disagreeable ... but the very sufficient reason, I fancy, is because the originals are so" (80). And as if to prove his point, Holgrave pulls out a daguerreotype miniature of the judge: "Now the remarkable point is, that the original wears, to the world's eye—and, for aught I know, to his most intimate friends—an exceedingly pleasant countenance, indicative of benevolence, openness of heart, sunny good humor, and other praise-worthy qualities of that cast. *The sun, as you see, tells quite another story"* (81) (my italics). In this, and in the final daguerreotype of Judge Pyncheon at his death, Hawthorne has discovered a resolution to his plot, a mechanism for breaking a spell of suspicion that has wasted most of Hepzibah and Clifford's lives. For in the judge's appearance at death, as recorded by the portrait, Holgrave discerns an exact likeness of the uncle whom Clifford supposedly murdered years ago; and he concludes that in both cases, the cause of death was natural and that Clifford had been framed by the judge himself.

It is a weak, implausible ending, but it is an ending that rests upon disclosure, upon the discovery of a truth long hidden and obscured by appearances. In a sense, it is a simple tale of the tables turned. For the judge, who had hoped to ruin Hepzibah and Clifford yet another time by discovering "the secrets of [the house's] interior," instead has his own deepest motives penetrated and revealed. To the extent that the ending works at all, it does so because Hawthorne has relied on current pho-tographic lore: "All nature shall paint herself." Since the daguerreotype was thought to be autotelic, since no individual sensibilities interfered with its rendering, its evidence was undoubtedly true.

Yet the problem that remained for Hawthorne was precisely that of "individual sensibilities," the problem of the photographer/artist whose own vision had been insufficient to the truth. In this case, the text is resolved by a daguerreotype: undisputed evidence is set forth so that Phoebe, Hepzibah, and Clifford can throw open the doors of their house and go unburdened into the world. But Hawthorne must have known that the camera was a gimmick, the displacement of a more central aesthetic and moral problem that was not confronted and resolved by the marriage, new house, and good fellowship that awaited his fictional family. He had managed "the lights and shadows" of his tale by a sleight of hand, for he must have known that the sensibility or vision of the camera and the photographer could not be separated as neatly as he had done, even though that separation was condoned by his contemporaries.

In all of his artist protagonists, Hawthorne found ways to dramatize his own position as a textmaker. If Holgrave had not been able to resolve the questions of truth that confronted him, but had had to rely on the camera, then what implications did that have for Hawthorne's own role outside the text? He claimed, as we know, to write from imag-ination. He had no intention of transcribing what he saw around him in literal detail. Nonetheless what did concern him in fiction—the truth of the human heart—depended on having knowledge of those truths from prior sources. In relation to them, was he the fallible daguerreotypist, snooping for clues but unable to read them correctly? Or was he the camera itself, equipped with both sight and insight, with the ability to accord proper weight to visual signs, to render light and shadow in unerring relation, to penetrate interiors that remained inscrutable to the common observer? As we know, he prided himself on his discernment, but in "Sights from a Steeple" his narrator had been foiled by surfaces, brought up short by his inability to see beneath exteriors. The resolution of *The House of the Seven Gables* depends upon that insight, but accords it to a machine.

Though Hawthorne never returned to fictions about cameras, it can be said that he never left his fictions about observers, about those who, like Holgrave, regarded the world with ulterior motives, with an eye to its usefulness to art: "It is not impulse, as regards these two individuals [Hepzibah and Clifford], either to help or hinder, but to look on, to analyze, to explain matters to myself, and to comprehend the drama which, for almost two hundred years, has been dragging its slow length over ground where you and I now tread" (175). If the common nineteenth-century currency was a language that detached the photographer both from the lives of his subjects and from the process of making photographs (nature recorded itself), Hawthorne had no illusion that he was uninvolved in the process of shaping fictions, and he knew that detachment from the lives of his subjects created its own set of problems. One could say that he wrote as if he were trying to play out the fiction that one could be just exactly what his contemporaries thought photographers to be: those who see without being seen, those who acquire knowledge without influencing the subject of that knowledge, those who live, for the purposes of taking portraits, outside the boundaries of reciprocal human intercourse.

IV

He repeated this obsession in *The Blithedale Romance.* There, in a utopian community, rather like Brook Farm, Hawthorne again dramatized the problem of witnessing lives and the difficulty of relating seer and seen. Instead of a daguerreotypist lurking with a camera, there is a New England poet who is accused of being on the Blithedale farm in order to "turn the affair into a ballad." In place of Holgrave's private desire for family revenge, Coverdale reveals another secret motive—one aside from his art—his undeclared love for a fellow community member,

Priscilla. Only in their ultimate fates do Holgrave and Coverdale differ substantially, since Holgrave eventually breaks his isolation and marries into the family he had so assiduously watched, while Coverdale faces the suicide of one of his "subjects" and the marriage of the other to his own rival. Hawthorne's later protagonist is left with the fruits of his own seclusion, just as the narrator in "Sights from a Steeple" had lost his secret attachment and had been thrown back on himself, successful neither in love nor in telling the story of love. Here it is Zenobia's death and the collapse of the community that severs Coverdale from the scene he wishes to analyze, but whether a storm or more ominous human conduct intervenes to stop the show, the result is the same anomie, the same desolation, the "old dull pain" that Longfellow so disliked as the mood of Hawthorne's writing.

The major structural difference between *Blithedale* and *The House of the Seven Gables* is a simple technical one. Where *The House of the Seven Gables* is a variation on a traditional third-person narration, *The Blithedale Romance* is told in the first person, with Coverdale's private, disturbed sensibility as the siphon of events. He is Hawthorne's oddest spokesman for the tensions between objective and subjective vision—at once the declared advocate of disinterest: "[I needed] to remove myself to a little distance, and take an exterior view of what we had all been about"[49]—and a predator who customarily makes "prey of people's individualities" (78). The implications of this changed narrative strategy are obvious: where Holgrave had been described, Coverdale explains himself as he suffers from the hostility his intrusiveness evokes. As if to justify his curiosity about the love lives of Zenobia, Priscilla, and Hollingsworth—the private network of relationships that underlies and in some senses undermines the utopian ideals of the community—Coverdale calls attention to duty, to his conviction that every human drama needs "the presence of at least one calm observer [who will] distill, in his long-brooding thought the whole morality of the performance" (90-91).

Here again it is possible to see Hawthorne setting up a dialectical situation for the artist, which has no potential for resolution: interest is always antithetical to clarity; engagement and understanding are mutually exclusive. Though Coverdale tries to justify his detachment, one could say that the text both condones that explanation and shows its delusions, for Hawthorne repeatedly undermines Coverdale's habitual posture by calling attention to his real, though unacknowledged, involvement in what he sees, and to the ethical and epistemological problems of uninvolvement. He loses either way. Zenobia, like Hepzibah in the previous novel, cannot bear scrutiny. She is more eloquent than Hawthorne's shriveled spinster, but the issue is the same: she sees observation as a violation: "I have often heard [the excuse of duty] before, from those who sought to interfere with me, and I know precisely what it signified. Bigotry; self-conceit; an insolent curiosity; a meddlesome temper; a cold-blooded criticism, founded on a shallow interpretation of half-perceptions" (157). Her rhetoric is hysterical, but it takes its place among a battery of intra-textual criticisms Hawthorne levels against his own methods. For he knew that the spectator could harm his subject by making it an object of investigation, just as he could cripple himself by extending an interest not wanted and even resisted. The veils and curtains that are literally interposed between Coverdale and his "objects" at the Blithedale farm are, of course, physical analogues of the psychological screens that the characters erect against Coverdale's "perceptive faculty." And in these screens, these active gestures of exclusion, we see a new aspect of Hawthorne's understanding of observation. Previously sight had been limited by what was available in the perceiver's field of vision: the artist was speculative, but ineffective in the world. Here sight occasions its own limitation. Watching actually creates the barrier that it was originally called on to penetrate. The result is manifested in the artist's repeated inability to tell the story "of the human heart" that he had wanted to tell. Without reciprocal communication there is no way for him either to confirm or deny the relation between outward expression and latent meaning. One can only presume a connection between an inner state and external appearance. As Zenobia warns, "[Yours] are only half-perceptions of half-truths."

These pained and convoluted issues of visual relationships stand in marked contrast to Hawthorne's other attempts at seeing the world. When we contrast both of these texts briefly with Hawthorne's notebooks, it becomes clear that his self-rebuke is integrally related to the process of making fictions out of experience. For in neither the *American Notebooks*—his comments about the natural world of Concord, Salem, Lenox, and his domestic life with Sophia Peabody—nor in his *English* and *Italian Notebooks*—the record of his travels abroad—do these issues of seer and seen arise. Alone, without the obligation of human intercourse, he is completely at home, appreciative, lucid. This serenity is all the more remarkable when we remember the anxiety that solitude had previously occasioned in Salem where he sat with his paper and pen as a young man, haunted by the spectres of his own imagination. *Twice Told Tales,* the fruit of those years, was meant "to open intercourse with the world," but his anxiety tells us how much that interaction was forced, the expression of ambivalence rather than of undivided desire. The woods offered him an alternative; they offered him no resistance, made no demands. Tuned to a cycle independent of human behavior, they offered themselves for scrutiny without problem.

Years later, when Hawthorne was thrust into a round of professional responsibilities by his consular appointment in England, he was similarly free of the anxiety of watching. Eventually assembled in *Our Old Home,* these sketches are the ones Hawthorne originally called "photographic," the ones he thought to be unimaginative and dull because he did not embroider or weave fancies around his experience. In "Outside Glimpses of English Poverty," he probably felt that the misery he saw was too consequential to be misrepresented. Yet if he imposed the obligation of fidelity on himself, he did not impose a new vantage point of observation. He was still, as the title of the sketch indicates, "outside," looking in on a subject poorly grasped, as always, aware of passing through life circumstances not his own, "a stranger . . . who might violate the filthy sanctity of the place." "There I caught glimpses of a people and a model of life that were comparatively new to my observation, a sort of sombred phantasmagoric spectacle, exceedingly undelightful to behold, yet involving a singular interest and even fascination in its ugliness."[50] Without commenting fully on Hawthorne's politics and his almost pathological inability to imagine that human effort might alleviate the suffering he sees ("I [see] the speedy necessity of a new Deluge . . . if every one of them could be drowned to-night"[51]), I would like to observe that Hawthorne attempts to deal with his own guilt by an act of identification. He is convinced that he must find a personal connection with the wretched people of London's slums: "What an intimate brotherhood is this in which we dwell. . . . It is but an example . . . of the innumerable and secret channels by which, at every moment of our lives, the flow and reflex of a common humanity pervade us all. How superficial are the niceties of such as pretend to keep aloof."[52]

The point is a simple one: Hawthorne had always been familiar with guilt, but where he previously had associated it with the act of seeing itself, here he circumvents that issue, or at least changes it into different terms. Since the gulf between London's poor and himself is caused by economic conditions that are beyond his control, since it is not caused by his own prying, Hawthorne is able to be a different kind of observer. One might argue that he is whistling in the dark, hoping against hope that a slum child's life, "my own life, and all our lives" have some connection in the sight of God, but in this context, such criticism is beside the point. For it is clear that despite his "loathsome interest" in the fetid living conditions of the poor, Hawthorne has lost his self-consciousness about watching, has become absorbed in the scene, his interest so rooted in what is "out there" that identification at least seems possible.

This freedom from guilt, a freedom similarly manifest in his nature studies, seems to come from a simple but crucial distinction: in these observations Hawthorne knows he is not using what he sees to his own advantage. His knowledge is not instrumental. In the case of making fictions, he could never forget that sight was a resource; the lives of those he observed were the useful prerequisites of his art, their character, behavior, values reduced to information on which he could perform imaginative transformations. Why else should he have wanted so desperately to become invisible? Had he wanted the human connection that he seems to want in this sketch on English poverty, he would have needed full visibility with all its requirements for reciprocal conduct. In short, the role of Paul Pry was dangerous to Hawthorne to the extent that it prohibited mutual relationship and turned lives into commodities for art. It was dangerous because it harmed both the artist in search of materials and those who were his subjects.

V

In closing, we must draw these observations about sight, insight, and the cost of vicarious experience back into the original inquiry about Hawthorne and photography. These connections are important because they show that a man who had very conventional attitudes toward the camera and the nature of photographic images was able to understand the process of taking pictures in ways that eluded most of his contemporaries. For Hawthorne, the daguerreotype represented simple truth, an extreme verisimilitude of surface detail and an unerring source of insight. Of these traits, he was interested only in the camera's capacity for insight. He saw "the inner man" as his true subject and chose freely among those fictional conventions that would allow him to render spiritual qualities most effectively. Photography never suggested itself to Hawthorne as a formal model to be emulated; he was not interested in the novel of concretized experience. What distinguished Hawthorne was his insight into photography as a human activity, for in his fears about predation, he acted as a precursor to many later artists who believed there to be more involved in taking pictures than arranging for the sun to be an artist. Through Holgrave, Hawthorne challenged his contemporaries' belief that photographs were artifacts acquired without a perceiving human subject and offered instead the opinion that photography as well as writing involved a human dynamic—the photographer's interest in the scene and the subject's response to that attention. His insight into the possibly damaging effects of vision derives from the combination of a private anxiety with a new technology; but this combination, however idiosyncratic in origin, anticipates the concerns of later generations who grew to be generally more sceptical of disengaged observation and more sensitive to intrusions made into human lives for the sake of art, science, or social progress.

These problems were not usually associated with photography in the mid-nineteenth century for a straight-forward reason: because of the technical requirements of making daguerreotypes, sitting for a portrait was almost invariably an act of consent. The subject came to a studio and asked for his portrait to be made. If he was self-conscious, he was at least not hostile. For Hawthorne, vision was never simple because it did not involve such consent; it had to be stolen. Between his wish that viewing could be a passive experience and his knowledge that it could not be, between his desire to be disinterested and his conviction that disinterest was impossible, Hawthorne located a set of problems, a web of ambivalences, that photographers and writers have faced with more and more clarity as time has passed.

Phoebe, in *The House of the Seven Gables,* articulated the problems of disinterest that haunt us still: "You seem to look at Hepzibah and Clifford's misfortunes ... as a tragedy ... played exclusively for your amusement. . . .The play costs the performers too much, and the audience is too coldhearted" (176). And Holgrave responded with its benefits: "Nor can anything be more curious than the vast discrepancy between portraits intended for engraving and the pencil sketches that pass from hand to hand behind the original's back" (223). Between these rival claims—of artistic truth and personal concern for those observed in the interest of truth—Hawthorne carved out an uneasy place for himself as an artist. "Realities keep in the rear," he said, and he was never convinced that anything but the "backyard" scene, the unposed gesture, could reveal the inner man. Yet he was never unburdened by the guilt of going unbidden behind the scenes.

Photographer unknown. From Nathaniel Hawthorne, *Transformation* (Leipzig: Bernhard Tauchnitz, 1860).

Photographer unknown. From Nathaniel Hawthorne, *Transformation* (Leipzig: Bernhard Tauchnitz, 1860).

Photographer unknown. From Nathaniel Hawthorne, *Transformation* (Leipzig: Bernhard Tauchnitz, 1860).

Alvin Langdon Coburn. The Curiosity Shop. London, 1906 *(International Museum of Photography at George Eastman House)*

2

Henry James and
Alvin Langdon Coburn
The Frame of Prevision

From science comes prevision, from prevision comes
control. —Auguste Comte

I

IN 1904, *Camera Work* carried a short story by one of its leading
contributors, Sadakichi Hartmann. It was a brief, unambitious piece
by a man with neither remarkable prose nor talent as a photographer—Hartmann's gifts were for criticism—but the tale catches a
moment familiar to photographers of his time: the search for the perfect image. In Hartmann's imagination, a photographer takes his sweet-
heart to an ocean resort and wanders over the countryside until her
beauty and the still brightness of the October day remind him of his
underlying mission: to capture such beauty on photographic plates, to
make his reputation as a photographer by the production of a master-
piece. Accordingly, the young man assembles his equipment and begins
to work feverishly. "Remain in that position!" he instructs the young
woman; and as he works, he loses track of her except as "a passing
shimmer, a flash of whiteness" in his composition. "Although she was
the center of all my enthusiasm, I seemed to have forgotten her actual
presence."[1]

The story turns on a simple accident: with the day's decline, his precious store of images packed in a case, the narrator helps his companion back to their hotel and drops his plates among the rocks. In retrospect, as he recounts the event, he seeks to express both an incalculable loss and his lover's callousness in the face of loss. Her feelings had changed; she no longer loved him; and his hopes of fame stood as shattered as the broken glass.

His current editor takes this story to reflect Hartmann's abandonment of artistic photography as a career. Whatever its biographical relevance, the sketch is interesting for broader reasons; for its assumptions, simple as they are, show the nature of photography sixty years after its introduction to this country, and they articulate the problems that photographers faced in establishing a place for themselves in the larger culture: Hartmann's ambitious and self-engrossed young man has moved out of the daguerreotype studio and into a new technology, a new realm of visual opportunities, and an attendant set of liabilities.

No longer burdened with awkward equipment, he carries his camera easily in the out-of-doors. Freed from lengthy exposures, he poses and reposes his model without the head-rest that had held daguerreotype subjects in stiff immobility. He makes images on glass plate negatives instead of metal. Given this technology, his immediate concerns are about "composition," about the "lines and values of the scene"; and his larger concern is to establish, in the same manner that more traditional artists do, a reputation through his work.

He is a character who would have been impossible to Hawthorne's imagination—for the obvious reason that he lives with technical advances unanticipated in the 1840s and for the equally important reason that Hawthorne and his contemporaries would never have considered photography to be a major art form. To them the camera's function and fascination lay in its ability to make accurate records. To the young man of Hartmann's story, "the contention . . . that nothing artistic could be produced by the camera filled me with indignation."[2]

That the protagonist's lover scoffs at his ambition tells us that photography, at that early moment of the century, had not firmly established itself as an art; and it seems no accident that the author of this story was at the heart of the debate about photography's status; a friend of Alfred Stieglitz, supporter of the Photo-Secession group; and an advocate of selected photographers whose work he considered to be as aesthetically valuable as painting. Those whom Hartmann admired varied in their approach to the print, differed on matters of composition, focus, degree of tonal contrast—in short, they did not agree about what formal qualities constituted photographic art—but all of them were opposed to utility and accuracy as the sole criteria of good photography.

Henry Peach Robinson can hardly be considered representative of the photographers whom Hartmann supported, but he did speak for them in one respect when he said, "Photography gives incomparably the greatest amount of power of minute imitation or copying with the most ridiculous ease, and it has lost the power of surprising us with its fidelity, for the detail of a photograph is one of the most ordinary objects of civilized life."[3] Robinson insisted that the inevitable likeness of photographs to the preexisting world need not exclude artistic arrangement of a print, that an "original interpretation of nature [was] ... limited ... but sufficient to stamp the impress of the author on certain works."[4] He amplified this view by explaining that the eye must be trained in order to recognize and frame the "accidental" beauties of nature and that every photographer, given the same subject, would produce different pictorial effects.

> My choice of the point of view, by the placing of a figure, by the selection of the time of day, or by over-exposure or underdevelopment, or by the reverse, producing soft, delicate, atmospheric effects or brilliant contrasts [lets me] render [my] interpretation of the scene, either as a dry matter-of-fact map of the view, or as a translation of the landscape so admirably suited to the subject.... Composition in art may be said to consist of the selection, arrangement and combination in a picture of the objects to be delineated.[5]

Robinson's personal rebellion against photography as a transcription of nature led him to manipulate the image, retouch negatives, and soften the focus of the lens in the service of "idealized beauty." He did not usually paint over the negative, but his tenets, so aggressively imitative of the more pictorial arts, led others to do so, and led, finally, to their own antithesis.

This challenge to art photography as championed by Robinson came primarily from Peter Henry Emerson who insisted that negatives should never be manipulated. But it is essential to notice that Emerson did not contradict Robinson's basic belief in photography as an art, but only Robinson's formula for composing certain kinds of artificial or sentimental images. He, too, believed that photography should be freed from its limited and limiting association with science, and he proposed, quite simply, that photographs be divided into categories determined by their use for either information or aesthetic pleasure. While he acknowledged the values of photographs whose clear definition, accuracy, and detail served the interests of science, his own inclinations were clearly aesthetic. Like Robinson before him, he insisted that "art is the selection, arrangement and recording of certain facts,"[6] and

though he denigrated Robinson's anecdotal genre scenes and opposed
tampering with the negative, it is clear that both men offered their
thoughts to the world in order to remove photography from its habitual
identification as a transcription or "plagiarism of nature" and to place
it instead in the category of art—as something premeditated and deeply
composed. Both of them challenged the public's original belief that the
form was autotelic and thus inevitably objective, and asserted instead
that the artist-photographer interfered with and shaped the final print—
either manually in the case of Robinson, or by virtue of a trained and
watchful sensibility. When describing photography at its inception,
Fox Talbot and his peers had insisted that the "sun was an artist."
Alphonse de Lamartine, after seeing some of Robinson's portraits in
1859, declared that photography "is more than an art, it is a solar phenomenon where the artist collaborates with the sun."[7] The historical
change marked by Lamartine's comment was clearly initiated by a new
perception of the role of the artist-photographer's shaping imagination,
by an end to the idea of vision as a passive activity, and by a sense of
the power inherent in observation itself.

All of these issues precede and explain the apparently simple decision of Hartmann's protagonist to compose his images and to become
deeply engrossed in "the lines and values of the scene" before him. He
is taking his stand as part of the avant-garde of his time. But it should
be noted that "The Broken Plates," although informed by historical
debate, is not about photography's controversial status among the arts,
but about the waning relationship of a man and a woman. The protagonist is incensed by his friend's lack of tenderness: "What more is there
to tell? Her indifference at that disastrous moment had deeply offended
me and gradually killed all my affection for her." Although the protagonist is mystified at her conduct, Hartmann writes with a fully
ironic awareness, for the grounds for understanding the woman's alienation are carefully laid. At the end of the afternoon, the ebullient photographer exclaims, "'But is it not wonderful that there is a whole
world around us to look at for years and years? And yet we are never
aware of it, we never see it until some happy moment suddenly reveals
it to us. And I owe it all to you!' 'By ignoring me,' she said,
reproachfully."[8]

At base, the vignette is a demonstration of one particular relation
between art and moral action. It is Hartmann's comment about the consequences of dissociated vision, and about the camera's way of promoting an instrumental view of reality. As the young woman of the story
clearly recognizes, the protagonist has made a choice that prohibits intimacy. She is of interest to her lover insofar as her beauty is useful.

Because she is not content to be "instrumental in [his] final success," his photography has occasioned the breach which he later finds so disconcerting. Hartmann's contribution to understanding photography as a mode of human interaction comes from this simple, but clean, demonstration of the camera's constitutive power to invade, to violate, and to create a set of dissonant relations in the very act which purports merely to record them. Where the photographer sees beauty, the subject of the photographer experiences, and acts on a distortion of motive occasioned by the aesthetic search itself. By focusing on the activity of the artist, the sketch dissociates photography from positivism with its supposed neutrality of stance, but it also demonstrates the fallacy of thinking the search for the picturesque to be an innocuous activity, devoid of its own manipulative strategies. Hartmann would have us understand that both modes of interaction—whether observation be in the service of science or of art—have a political status. Because a relation of knowledge is, for him, a relation of force, he considers photography to be enmeshed in, and emblematic of, a series of complicated social structures.

In this one respect Hartmann's sketch is of pivotal importance: it looks backward and forward in time, reminding us of Hawthorne's dilemma in "Sights from a Steeple," where the very act of watching a young woman had prohibited the narrator from making her acquaintance. And it anticipates, as I will demonstrate, Henry James's poignantly informed debate about "the expense of vision" in his fiction. Like Hartmann, James was acutely and anxiously aware of the constellation of aesthetic and political issues occasioned by his own spectatorship.

II

A year after Hartmann's story was published, in April 1905, Henry James met Alvin Langdon Coburn. Coburn was then twenty-three, at the beginning of his career as a remarkable portrait artist, and he had requested a sitting from the noted author for a series of photographs of literary celebrities to run in *Century Magazine.* "And so it all began," Coburn recalled in a B.B.C. broadcast long after James's death.[9] For what was a casual meeting grew by chance into a lengthy collaboration between writer and photographer, a collaboration augmented and directed by over fifty letters James wrote as he strove to previsualize the photographic frontispieces for the New York edition of his work, to articulate the vague discomfort that had led him to eschew the more

traditional kinds of book illustration, and to work through the theoretical relation between photography and fiction.

Their work together started with a simple invitation. At his first
sitting in America, James had learned that he and Coburn were both
expatriates, a coincidence which pleased him and led him to ask Coburn to Rye upon their return to England. He wanted, he had decided,
a portrait for the frontispiece of his collected works. On 12 July 1906,
Coburn came as requested to take the picture, and two weeks later he
was asked to return to Lamb House. "It is now important that you
should do my (this) little house (for the same use) and could you come
down for the purpose ... some day early next week? Tuesday or
Wednesday?"[10]

We do not know when James decided that the single portrait
should initiate an entire series of photographic frontispieces; but in the
preface to *The Golden Bowl,* he did leave a clear account of how the
thought came to him. Here, amid all his self-reflection and retrospective evaluations, he spoke of photography as antithetical to fiction. It
was, as he saw it, the least competitive form of visual representation,
and since he thought good narrative to consist of vivid images, this was
very important to him:

> [T]he proposed photographic studies were to seek the way, which
> they have happily found, I think not to keep, or to pretend to keep,
> anything like dramatic step with their suggestive matter. This would
> quite have disqualified them, to my rigour; but they were "all right,"
> in the so analytic modern critical phrase, through their discreetly
> disavowing emulation.[11]

His writing should stand on its own merits; he would have no other
artist conceptualize or seek to represent that which language alone
should conjure. In his imagination these frontispieces assumed the function of a stage set. Devoid of people, they should suggest, in the most
general way, the physical arena on which his characters would subsequently act out their fates. They would be only "images always confessing themselves mere optical symbols or echoes, expressions of no
particular thing in the text, but only of the type or idea of this or that
thing." He gave as an example the problem of finding the proper photograph for *The Golden Bowl.* Having decided that nothing would be as
suitable as a view of the antique shop in which the bowl was first
encountered, he and Coburn set out to find in the real world that which
had existed previously as a "shop of the mind," "the small shop ... of
the author's projected world, in which objects are primarily related to

each other, and therefore not 'taken from' a particular establishment anywhere."

James accompanied Coburn on these hunting expeditions, searching out the physical spots that would accord with his own imaginative vision. The London world became a set of recalcitrant facts that might, by prodigious effort and a trained eye, be induced to yield "the aspect of things or the combination of objects that might, by a latent virtue in it, speak for its connection with something in the book."[12] He also wanted the photograph to be excellent for its own sake—as he put it, "to speak enough for its odd or interesting self," but at no time did he equate the photographer's activities of discovery, selection, and isolation with his own art. In his mind, the antithesis between photography and fiction continued—the photographer had only to recognize but not create. Six years after he began work with Coburn, he was still asserting, "Photography insists for me, in remaining at best *but* photography."[13] With all the fondness and admiration he apparently had for the young and talented Coburn, James did not see him as a professional rival, but almost as a servant who could act, with proper instructions, as an extension of the writer's mind. James speaks of the world becoming malleable enough for *him,* the writer, to use, of inducing it "to generalize itself." We must remember that before Coburn met James, he had become a member of the Photo-Secession group in the United States and had been elected to the "Linked Ring" in London. In 1906, he had exhibited at the Royal Photographic Society—his second one-man show in London in six years. If any were devoted to establishing photography as a fine art during the early years of this century, Coburn was certainly among them. Yet James did not fully come to terms with the implications of his young friend's activities. Instead he adhered to a notion of the photographer's essential passivity or blankness and the photograph's salient quality as a transcription of nature which could, in this particular instance, be made consonant with a previsualized imaginative place.

With as much thought as James apparently gave to this collaboration, he continued to conceptualize the process of photography as he had in the years previous to meeting Coburn; that is, he seems to have retained the view that had been current in his youth and aligned it primarily with positivism. James's fiction, replete as it is with conscious analogies to the other visual arts, gives scant attention to the camera. The aristocratic woman who sits for the artist in "The Real Thing" sits stiffly and self-consciously—as if in front of a lens—and the painter, influenced by her rigidity, begins to behave like a camera and to copy her outline rather than interpret her character. "But after a few times I began to find her too insurmountably stiff; do what I would with it my drawing looked like a photograph or a copy of a photograph."[14] The narrator in *The Aspern Papers* rues the advent of photography and considers it to be instrumental in destroying the freshness of Europe because the print could anticipate places without their discovery in experience and because that anticipation promoted, as he saw it, overwhelming, uninformed, and insensitive tourism. "When Americans went abroad in 1820 there was something romantic, almost heroic in it, as compared with the perpetual ferryings of the present hour, when photography and other conveniences have annihilated surprise."[15]

Like Ralph Waldo Emerson, James thought the camera to be unable to capture the vitality of life; and unlike Oliver Wendell Holmes, who relished "armchair travel," he denigrated the vicarious visual experience offered by pictures. Although he was often photographed and although he was repeatedly delighted with Coburn's work, he himself was satisfied with nothing less than the real view, the scene with all the tangible grace of atmosphere that photographs inevitably missed.

All of these were notable views for James to hold, especially in light of Alvin Langdon Coburn's particular genius and artistic propensities, for they were not at all consistent with the way that Coburn saw himself in 1905. In fact, the two men seem to have espoused the categories Peter Henry Emerson had extended, each beginning with opposite assumptions about the camera. Where James aligned photography with science because it was an indiscriminate recorder of fact, Coburn of course considered it a fine art. He had dedicated his life to it early, completely, and without hesitation. The daguerreotypists of Hawthorne's time may have been a vagabond lot, but Coburn exhibited none of their wanderlust, none of their dextrous juggling of skills. He was a photographer as simply and completely as he could be, and had been since his youth. Sadakichi Hartmann tells us that Coburn's first teacher was his own cousin, F. Holland Day, in whose studio he "mastered the technical problems of his profession"[16] and that he later studied with Gertrude Kasebier and with Arthur W. Dow, the American landscape painter. He first exhibited in London in 1900, and by 1904 was known widely enough to participate in one of the first major Photo Secession exhibits. From this early point in his life, Coburn was associated with the Photo Secession group—united with them in their rebellion against the manipulated print, insistent that the camera must record what originally existed in nature, and intent nonetheless on creating beauty through the careful choice of "frames." During the period when James

knew him, Coburn divided his energy between portraiture and landscape photography, but in neither activity did he adhere to sharp resolution or rendering of fine detail. He was concerned instead with the
harmony of elements within the frame of the print, with rendering,
through adjustment of forms and attention to light and shade, a subjective sense of the life before his lens. Photography as Coburn understood
it was highly expressive and inevitably bore the mark of the photographer's personality. His aim, he wrote with regard to landscape photography, was "always to convey a mood and not to impart local
information."

> You ask how a camera can be made to convey a mood, I can only say
> that photography demands great patience: waiting for the right hour,
> the right moment, and recognizing it when you see it. It also means
> a training in self control. . . . The artist-photographer must be con-
> stantly on the alert for the perfect moment, when a fragment of the
> jumble of nature is isolated by the conditions of light or atmosphere,
> until every detail is just right.

> To speak of composition in connection with photography seems, on
> first thinking of the problem, to be rather a contradiction in terms—
> that you really ought to say "isolation" which would perhaps come
> nearer to what is done in most cases; but whilst it is impossible to
> rearrange trees and hills in the manner of the painter, it *is* possible to
> move the camera in such a way that a completely new arrangement
> is achieved, a few inches sometimes changing the entire design. For
> the creation of a picture, vision is of prime importance, and patience,
> discrimination, and even marksmanship are decisive factors.[17]

In these several remarks we can see Coburn sorting out both the
limitations and subjective possibilities of his medium. He understood
photographic creativity to be a series of changes worked upon given
material which could not be changed in itself. The photographer had
to take what was "there"—the trees and hills could not be rearranged—
and use reality as a basis for invention. But invention, in his sense,
consisted of a highly developed, acutely informed receptivity. Coburn
would have agreed with James that photographers had "only to recognize," but instead of opposing recognition to creation, he would have
identified recognition or informed vision as an active talent. In this
distinction lay the fundamental disparity between the two men's views
of photography—and again it was a curious disagreement for them to
have, with regard to his own work, James struggled openly to justify the same things Coburn did: he spoke of the novelist's life of vision
as an active and viable mode of being and he also knew full well the

determining role of the angle of observation. Where Robinson talked of "point of sight" and Coburn emphasized "angle of vision" or the importance of the frame to the total composition, James explained the formal qualities of his narrative in terms of "point of view" and revealed the process of gathering materials for narrative to be, at least at times, analogous to the educated and poised moment in which the pictorial qualities or loveliness of nature should be revealed. "Beauty is always there to be found by the trained eye," Coburn had written and had added that "one of the values of photography is the training it gives in the discovery of what constitutes a picture."[18] Quite independently, James had observed, "Though the relations of a human figure or a social occurrence are what make such objects interesting, they also make them difficult to isolate, to surround with the sharp black line, to frame in the square, the circle, the charming oval that helps any arrangement of objects to become a picture."[19] He also identified the novelist as a "seeking fabulist [whose] discoveries [were] scarce more than alert recognitions,"[20] and spoke about realism as a commitment to write about "life without rearrangement."

Even as they are briefly juxtaposed here, the parallels in these descriptions are too insistent to overlook—trained observation, the lurking sensibility with the talent for framing and isolating a picture, the commitment to representation of the unarranged world—all of these coincidences tell us not that James consciously imitated the camera—for he did not—but that similarities of procedure and commitment existed beyond his readiness to see them.

III

James's criticism of photography, or at least his sense of its limited possibilities, derived, then, from his preconception of the camera as an indiscriminate recorder of details. That Coburn and many of his contemporaries had moved to another, almost antithetical sense of the uses and procedures of photography is one of the unmarked ironies of the situation the two men found themselves in. For James's struggle was in many ways the same as Coburn's—to define and practice an art that avoided indiscriminate recording, that achieved formal coherence without encyclopedism. His critique of the camera was, then, essentially the same as his argument with the French naturalists who worked by inclusiveness rather than selection and who were, he thought, victims of their own exhaustively documentary methods.

James came to terms with these particular novelists—Balzac, Flaubert, Zola, de Maupassant—by writing about them. On his visits to Paris he had met a number of them; he had enjoyed Sunday afternoons at Flaubert's home and evenings with Zola and Concourt at the invitation of Alphonse Daudet (1884). Yet it was not conversation or personal friendship that James sought with these men, but views on how to write. That he rarely endorsed their thoughts was inconsequential, for his own identity as a writer was honed on the edge of disagreement. He was not indiscriminate, nor was he close-minded. The novelists I have grouped together existed independently for him; each offered him something unique. Nonetheless a common strain runs through James's criticism of them, and his dissatisfaction in all cases has to do with a mode of observing and notating experience which one could call camera-acting. Speaking of Balzac, the French photographer, Nadar, had observed, "The procedure of photography is a materializing so to speak of what is most original in his procedure as a novelist."[21] And Zola, following Balzac's example, had remarked, "The observer sets down purely and simply the phenomena he has before his eyes. . . . He ought to be the photographer of phenomena; his observation ought to represent nature exactly."[22]

One can argue about the conscious influence photography exerted on the works of either of these men—it is an intriguing speculation, since Balzac was a friend of Nadar and Zola was an avid amateur photographer—but in neither case is it possible to dispute a method of observing and interacting with the world that, if not promoted by the camera, was at least shared by it: naturalistic novelists and scientific photographers were all dedicated to precise and impartial specification of the material world.

Balzac, as James saw him, was a master of the robust, epic novel, the novel built on a full catalogue of details about persons and milieux. He was indebted to Balzac, and yet, finally, critical of his appetite for proliferation. In "The New Novel," James wrote disparagingly of "that comparative desert of the inselective, overflooded surface," of "undiscriminated quantity," of "saturation of detail" as opposed to an "interesting use of material." His final view of the master included the objection that "he sees and presents too many facts"—a criticism he also extended to de Maupassant and Zola.[23]

With the latter writers, James argued about choice of subject—he thought the decision to write primarily about lust and poverty as narrow as his predecessors' determination to keep passion out of their work—but he did not harp on his point. It was his belief that one should grant every writer his donnée and ask only what one had done with the given material. And again, with regard to technique, his criticisms were about multiplication, accumulation, mechanistic vision, and the impossibility of an author assuming the role of an impartial, non-participating observer of his own fictional world. Of Zola's *Vérité* he mused, "Machine-minted and made good by an immense expertness, it yet makes us ask how, for disinterested observation and perception, the writer had used so much time and so much acquisition, and how he can all along have handled so much material without some larger subjective consequence."[24] He was fascinated by Zola's monkish existence, his lack of experience, and his consequent reliance on documents for material. "Poor, uninstructed, unacquainted, unintroduced, he set up his subject wholly from the outside, proposing to himself . . . to get into it, its depths, as he went." He concluded that Zola had let "breadth and energy supply the place of penetration."[25]

Of de Maupassant's rationale for the outside view—that novels should avoid "all complicated explanations, all dissertations upon motives, and confine [themselves] to making persons and events pass before our eyes"—James was similarly sceptical: one could not describe an action without some notion of its motive; one could not describe anything at all without a personal, thus partial, opinion of it. "If a picture, a tale, or a novel be a direct impression of life," he claimed, "the impression will vary according to the plate that takes it."[26]

To the extent that these novelists were interested in the camera or influenced by it, their interest was manifested by using the techniques of objective observation that photography offered to them as models: they would stand before their subject as the scientific photographer before the world, impassive, recording the full panorama as if their own eyes were no more than mechanical lenses. The interior world of psychological motive, which was necessarily hidden or unavailable to the camera eye, they would eschew by choice. These very decisions are what James rued, and his essays on the naturalists, written comparatively early in his career and before the accomplishment of most of his own major fiction, reveal the grounds for his own rejections and reformulations of the novelist's task. Where the naturalists worked by the juxtaposition of items, "emulating . . . a schoolboy's sum . . . a tiresome procession of would-be narrative items in addition," he would uphold "the mystery of the foreshortened procession of facts and figures,"[27] letting the carefully selected glimpse speak for the whole. Where the naturalists respected the unarranged world and represented it by including both banal and momentous details, he would make narrative pictures "governed by the principle of composition." He would use discrimination in selecting and arranging material according to a consistent point of view. The novelist should be unobtrusive; he should work to eliminate the appearance of a manipulating author—in this James was in agreement with Flaubert and Zola—but he thought it pointless to deny that some hidden sensibility shaped the final narrative.

The irony in all of this is of course that these tenets were precisely the aesthetic strictures advocated by the later pictorialist photographers—themselves in revolt against positivism's expropriation of the camera. One hears the echo of Emerson, Stieglitz, Coburn in all of these narrative principles, and, in fact, Coburn's description of his aims and procedures is as beautiful a way of explaining James's aesthetic philosophy as I know: the artist-photographer worked with a world that was given (James's "life without rearrangement"), "moving the camera in such a way that a completely new arrangement is achieved, a few inches sometimes changing the entire design."[28] Realism for James was always a modulation of allegiances to the existing world, to the "direct impression," and to the imagination which recognized and framed, not by means of unbridled inventiveness, but by changing raw data into composition through perspective.

Perhaps it was this fundamental though unacknowledged agreement in aesthetic aims that explains the enjoyment both James and Coburn derived from their collaboration. Although the novelist's Golden Bowl preface carefully explained that he and photographers worked in antithetical modes, his own theories tended to soften and qualify this distinction, as did the letters of instruction he wrote to Coburn as their work together progressed. These letters show that each photograph of the eventual series of frontispieces originated with James's prevision of the final image. "I peruse your List," he wrote to the young photographer, "but we are not right yet about all the Subjects, and it takes a good deal of worrying out, as I find face to face with all the things I have to combine and reconcile and fit in."[29] Years later, looking back on their years of association, Coburn was led to observe:

> [A]lthough [he was] not literally a photographer, I believe Henry James must have had sensitive plates in his brain on which to record his impressions! He always knew exactly what he wanted, although many of the pictures were but images in his mind and imagination, and what we did was to browse diligently until we found such a subject.[30]

The search for frontispieces took the younger artist to three continents over a period of several years. When the appropriate setting of a text was London, the two men wandered about the city together, James

as guide, Coburn, according to both their accounts, as the willing
instrument of James's intentions. In his own memories of these afternoons, James wrote about their passive watchfulness, as if reality would
arrange itself at the perfect moment and pause for its capture on the
plate:

> The thing was to induce the vision of Portland Place to generalize
> itself. This is precisely, however, the fashion after which the prodi-
> gious city, as I have called it, does on occasion meet halfway those
> forms of intelligence of it that it recognizes. All of which means that
> at a given moment the great Philistine vista would itself perform a
> miracle, would become interesting for a splendid atmospheric hour,
> as only London knows how, and that our business would be to
> understand.[31]

And yet, that poised receptivity worked always in tandem with an
anticipatory interpretation of observed phenomena. James's watchfulness was informed by deep premeditation about the image wanted. As
Coburn went on assignment to Paris, Venice, Rome, and America, he
carried with him James's account of the type of picture that should be
taken. At the Place de la Concorde James advised:

> Look out *there* for some combination of objects that won't be hack-
> neyed and commonplace and panoramic; some fountain or statue or
> balustrade or vista or suggestion (of some damnable sort or other) that
> will serve in connection with *The Ambassadors,* perhaps; just as some
> view, rightly arrived at, of Notre-Dame would also serve—if suffi-
> ciently bedimmed and refined and glorified; especially as to its Side
> on the River and Back ditto.[32]

For *The Aspern Papers* he wanted a picture of Juliana's Court. "I have
just written to Miss Constance Fletcher, in Venice where she lives. . . .
I have told her exactly what I want you to do."[33]

> The extremely tortuous and complicated walk—taking Piazza San
> Marco as a starting point—will show you so much, so many bits and
> odds and ends, such a revel of Venetian picturesqueness, that I advise
> your doing it on foot as much as possible. . . . It is the old faded pink-
> faced battered-looking and quite homely and plain (as things go in
> Venice) old Palazzino on the right of the small Canal, a little way
> along, as you enter it by the end of the Canal towards the Station. It
> has a garden behind it; it doesn't moreover bathe its steps, if I remem-
> ber right, directly in the Canal, but has a small paved Riva or footway
> in front of it, and *then* water-steps down from this little quay. As to
> that, however, the time since I have seen it may muddle me; but I
> am almost sure. At any rate anyone about will identify for you Ca

Capello, which is familiar for Casa C. . . . You must judge for your-self, face to face with the object, how much, on the spot, it seems to lend itself to a picture. I think it *must,* more or less, or sufficiently; with or without such adjuncts of the rest of the scene (from the back opposite, from the bank near, or from wherever you can damnably manage it) as may seem to contribute or complete—to be needed, in short, for the interesting effect. . . . What figures most is the big old Sala, the large central hall of the principal floor of the house, to which they (my friends) will introduce you, and from which, from the larger, rather bare Venetian perspective of which, and preferably looking towards the garden-end, I very much hope some result. In one way or another, in fine, it seems to me it ought to give something. If it doesn't even with the help of more of the little canal-view etc., yield satisfaction, wander about until you find something that looks sufficiently like it, some old second-rate palace on a by-canal, with a Riva in front, and if any such takes you at all, do it at a venture, as a possibly better alternative. But get the Sala at Ca Capello, without fail, if *it* proves at all manageable or effective.[34]

In James's advice and prodding one sees his tendency to use Coburn's camera vicariously, as if it were an extension of his own imagination—the subject, the view, the angle, the mood anticipated in advance of Coburn's expedition, his own previous knowledge of places used as the basis for the cameraman's work of framing and recording. As a proce-dure, it is reminiscent of Stieglitz, who was described as "fully conceiv[ing] his picture before he attempts to take it, seeking for effects of vivid actuality and reducing the final record to its simplest terms of expression."[35] And yet the analogy also seems to extend to James's own work and method of conducting life in preparation for work: experi-ence leading him to the "story," the story achieving validity by adher-ing to the actual scene, but achieving formal coherence through a set of imaginative transformations. Structure emerged as a function of focus as the artist's hidden but active sensibility located and emphasized aesthetic relations that existed in the midst of undifferentiated and undistinguished details.

The analogy should not be pushed too far, but this juxtaposition allows us to see that James participated in a debate about formal aes-thetics that was wider than literature and that involved both photog-raphers and novelists in decisions about the imagination's interference with or rearrangement of reality. For the naturalists, who, James thought, imitated the camera as it was used in the service of science, objectivity was of paramount importance. They wrote as if they could obliterate their own sensibilities or as if the scene before them would somehow be diminished by their subjective appraisal of it. Giovanni Verga wrote, "The hand of the artist will remain absolutely invisible, then it will have the imprint of an actual happening; the work of art will seem *to have made itself,* to have matured and come into being spontaneously, like a fact of nature, without retaining any point of contact with its author."[36] For James and Coburn, however, the shaping imagination was both an inevitable and a desired aspect of their craft. They both retained an elitism of vision; and in this elitism and their recognition of what it implied about the social dimension of artistic creation, is contained one of their most distinguished contributions to our understanding of realism and its dependence upon the observational structure of positivistic science. Their art is simultaneously an engagement with, and a criticism of, the ideology of objectivity. It shows us a fallacy: when Verga spoke of the work of art making itself, when he stressed the neutrality of the recording artist, he aligned a literary movement and a mode of artistic creation with another investigative discourse that gave priority to scientific method, and acknowledged only it as valid. Both James and Coburn knew that their successes rested on the kind of highly trained sight that isolated and revealed the dramatic moment. No matter how hidden they might seem to be, they understood that vision could not be uninflected. Their sensibility interfered in their art, and it interfered in the lives of their subjects as they worked actively, like Robinson before them, "in collaboration with the sun."

IV

Although James and Coburn were very close when they were working together, their attitudes toward their subjects were ultimately quite different. With landscape photography of the kind James wanted for his frontispieces, this divergence was unimportant; but both men were deeply and, I think, primarily concerned with live subjects—James, of course, as a fabricator of human fates, Coburn as an avid portrait photographer. Coburn's *Men of Mark* series represents many years spent in pursuit of public personalities who represented the achievements of the age. In a modest and unassuming way, he was a hero-worshipper who pre-arranged sittings, approached his chosen subjects with consistent respect, and tried to elicit their characteristic greatness: "The camera naturally records the slightest change of expression and mood and the impression that I make on my sitter is as important as the effect he has on me."[37]

James, too, was sensitive to the reciprocal influence of artist and subject, but for him the terms of his visual access to the world were much more problematic. Coburn requested sittings, but James could not ask his subjects for a similar access to their lives. Then, too, his moments of human contact were not, like Coburn's, his moments of creativity, but only the prerequisite for narratives that would be formulated in contemplative retrospection. These distinctions led James to experience tensions that Coburn never felt: tension between gathering useful materials and writing, and between the impulse to use the world as potential material and a conviction that social interaction should not be instrumental. The first of these anxieties was expressed in his essay on Balzac, but it was a dilemma which applied, as he knew, with equal relevance to himself:

> He could live at large, in short, because he was always living in the particular necessity, the particular intended connection—was always astride of his imagination, always charging, with his heavy, his heroic lance in rest, at every object that sprang up in his path. . . . But as he was at the same time always fencing himself in against the personal adventure, the personal experience, in order to preserve himself for converting it into history, how did experience, in the immediate sense, still get itself saved? Or to put it as simply as possible, where, with so strenuous a conception of the use of material, was material itself so strenuously quarried?[38]

James's second anxiety, a continuation of Hawthorne's guilt about prying and Hartmann's bruised sense of his own estrangement, was expressed throughout his fiction. "Where," James had asked of Balzac, "with so strenuous a conception of the use of material, was material itself so strenuously quarried?" It was a question he knew to ask because he himself was accustomed to amassing such material and because he was well aware of the cost of his appropriating impulse. James did not speculate about the personal loss involved in Balzac's writing career, but he did describe himself in terms reminiscent of the photographic sketch that opens this chapter: just as Hartmann's protagonist had faced grudgingly the penalty exacted by his camera work, James came to understand the price of writing novels in terms of dissociated vision. He saw himself engaged in a mode of life that committed him constantly to noticing rather than participating. Observation, "quarrying" material, and the separation or alienation that watchfulness entailed, all came together for him, as they had for Hawthorne, in terms photography could aptly describe: vision became a predatory activity—one which James spent many years covertly describing and ameliorating in his novels.

In one respect, the problems of observation were at the core of all of James's fiction. Looking back on his achievements in 1907, as he wrote the last preface for the New York edition of his work, he was able to identify one of his major technical devices—point of view—as both a formal narrative technique and as an indirect method of self-dramatization. It was, he explained, a way of discussing within fiction the problems that he faced in that fiction's creation. He spoke of "'seeing my story' through the opportunity and the sensibility of some more or less detached, some not strictly involved, though thoroughly interested and intelligent witness or reporter"—a person who was or could be an "unnamed, unintroduced and (save by right of intrinsic wit) unwarranted participant," and, most significantly, someone who would thus serve as a "a convenient substitute or apologist for the creative power otherwise so veiled and disembodied."[39]

The significance of James's remark consists largely in his retrospective reflection that, for him, techniques of vision had often provided both the method and the subject of his narratives, that form and substance had often combined as a substructure of interest beneath or in addition to his more obvious themes. But though this impulse for conscious meta-fiction was a continuous one, a few of his novels focus more clearly than do others on the moral and epistemological implications of his decision to lead a life that was psychically dependent on the behavior of those observed. This dependence was initially perceived as a solitary and private liability, but eventually James understood his dilemma to have wider social significance. The novel that most clearly illustrates James's discomfort on the private level is *The Sacred Fount;* the novel that dramatizes his expanded awareness is *The Princess Casamassima.* The first text, which limits itself to the petty power plays among the idle rich, nonetheless lays the foundation for understanding the ways in which psychological and social/economic structures can mimic each other, each reiterating a similar mode of visual negotiation. Both novels explore the relation between seeing and power; or, as Mark Seltzer has said so cogently, their narrative techniques enact a power play that is, on other levels of textual analysis, also the overt subject of inquiry.[40]

The Sacred Fount is a book whose interest lies partly in its unsatisfactory plot; for the broken nature of the story calls attention away from the overt line of events and back to the structure that underlies and empowers it. It is a book which is both indeterminate in meaning and yet completely in James's control, and its most pressing concern is aesthetic method. In it James asks what are the limits and culpabilities of knowledge gained without the consent of the subjects of that knowledge. What does an asymmetrical relation of seeing tell us about the nature of domination and repression? He then enacts those limitations and problems through an unnamed narrator who, like Hawthorne's narrator in "Sights from a Steeple," cannot deliver the story he had initially hoped to tell because he cannot gain definitive access to the requisite material. His subjects' refusal to cooperate in this acquisitive venture then shows the extent to which they feel repressed by his spying.

Ostensibly the novel is about a weekend party at an English country house. Three of the guests encounter each other on the train to Newmarch, and as they chat, the narrator notices a change in both of his companions. Mrs. Brissenden, a plain, middle-aged woman, seems unusually vivacious, and Mr. Long is more articulate and witty than the narrator had remembered. As he ponders these circumstances, he recalls that Mrs. Briss has recently married; but rather than deciding that marriage agrees with her, he formulates a more intricate explanation: Mrs. Briss has drawn her energy and restored youthfulness from her husband who has aged in the process of nourishing her. Then, delighting in the symmetry of this waxing and waning, he extends the theory to explain Long's unwonted cleverness: to complete the syllogism, there must be a woman whose intelligence Long is draining. Of the Brissendens he says, "One of the pair ... has to pay for the other ... he on his side, to supply her, has had to tap the sacred fount."[41] He is also convinced that someone is similarly victimized by Long.

Whatever charms and entertainments the weekend holds for the other Newmarch guests, the party becomes a quite different affair for the narrator. Wandering among his companions, he seeks first to discover Long's secret lover and then to find definite proof of their liaison. His conversations, his private reflections, his covert observations are all directed toward this end, and amid all of James's annoying refusals to clarify the case, to tell us whether these lovers do exist in the projected relationship, he never wavers in giving evidence that the narrator has changed his own status within the group: he is no longer a participant but a witness. The reality of Newmarch, with all of its social intricacies and elaborate decorum, has become a target for his surveillance.

The play between the narrator, the Brissendens, Gilbert Long, and May Server consequently turns from pleasantries to proof: their behavior is used by the narrator to confirm or deny his dark, contrived theory of human intimacy; and as the weekend progresses, the others react, each in a characteristic way, to his excessive interest in them. While they group and regroup themselves for casual talk in the library, strolls on the lawn, seating at dinner, the protagonist keeps a hawk-eye on their activities, postulating both intimacy and a desire to hide intimacy in what he sees ("as the appearance is inevitably a kind of betrayal, it's in somebody's interest to conceal it" [36]). He trusts neither appearances nor confessions since both could be construed as screens or purposeful obstructions to the truth.

To live in this strangely poised manner—interested but outside, intellectually engaged but ignorant—has its price, and it is in the evaluation of that price that James writes with clearest dramatic irony, for he lets us see losses beyond his character's subjective appraisal of them.

His narrator considers that he must be objective in approaching his self-generated problem: "The condition of light," he remarks, "was in this direct way, the sacrifice of feeling" (14). Like Coverdale in *The Blithedale Romance,* he considers understanding to be antithetical to involvement. For him, truth becomes a function of nonsubjectivity, and in the interests of truth, he reduces human relations to collectable "data." These self-imposed rules proliferate, but in the end they reduce to a privately applied empirical method—the narrator has bound himself to make deductions from only the observed or "scientifically" documentable aspects of experience. And what he learns by this method, as James shows us, is sparse: in fact nothing can be verified about May Server's or Gilbert Long's love life. None of the other guests agree about the Server/Long relationship, and as the weekend wears on, those who know of the narrator's theory pass from amusement to incredulity and then to an uncomfortable sense of proprieties overstepped. Those who had playfully participated in his speculations back out. Long, who had once agreed that Mrs. Briss was rejuvenated, now insists that she is plain and middle-aged. The artist, Mr. Obert, concludes that May Server has "changed back" to her old self and suggests that "success in such an inquiry may perhaps be more embarrassing than failure. To nose about for a relation that a lady has her reasons for keeping secret [is appalling]" (57). Finally Grace Brissenden arranges an evening confrontation to tell the narrator bluntly that he is crazy, that his scrutiny of everyone is terrible. To set his mind at rest and, she hopes, to end his meddling, she tells him that her husband, a confidant of Long, knows Long to be having an affair with Lady John, and that May Server has been making love to him (Briss).

Of course none of this information is conclusive since it could once again fall into the category of purposefully planted screens—Mrs. Briss may be trying to mislead the narrator—but the point is really that such screens are perceived to be necessary, that the narrator has identified himself as a kind of psychological leech who interests himself, for the sake of a theory, in what is none of his business. By translating his friends into functions of analytic discourse, he has moved from parity with them to a power relationship. Upon reflection, we see that the whole situation of the novel—the country setting, the closed society thrown into intimacy, the decision of one group member to extricate himself from the group for the sake of "impartial" observation, his observation resisted—resolves itself in much the same way as Hawthorne's *The Blithedale Romance*. Once again, scrutiny occasions its own limitation, for it destroys a balance, obliterates full reciprocity of emotion, and establishes a dim, uncomfortable sense of social interaction undertaken with an ulterior motive. Gazing is interpreted to be a kind of power play which generates an increased desire for secrecy on the part of those observed: in response to scrutiny they impose a psychological repression upon themselves, a repression which constricts their own behavior, but which also, by its exclusiveness, attempts to restore a measure of self-control. This is a circumstance the narrator seems already to understand: "And I made a final induction. The agents of the sacrifice are uncomfortable, I gather, when they suspect or fear that you see" (35).

The irony of the story derives, then, from a clean and almost total reversal: where the narrator had hoped to prove a theory about "vampires" in intimate love relations, he himself demonstrates that same propensity to prey on others. He participates, vis-à-vis the objects of his contemplation, in the same waxing and waning that he had hoped to discover in the outside world. Indeed, when one looks back over the garden party with all of the narrator's forays into calculated sociability, one sees how James establishes this reversal by using the language of diminishment and gain and by calling attention to predation. "One of the pair," the narrator had concluded about the Brissendens, "has to pay for the other," but as time progresses, he himself pays or gains, according to the success of his surveillance: when Guy Brissenden shows up, the narrator admits "his being there . . . renewed my sources and replenished my current—spoke all, in short, for my gain" (157). Brissenden himself perceives this relationship when he rejoins that the narrator must be happy "a little so now at my expense" (93). On the other hand, the narrator sees that his obsession has an underside: "It was absurd to have consented to such immersion, intellectually speaking, in the affairs of other people. One had always affairs of one's own, and I was positively neglecting mine" (72).

This teetering back and forth continues, the narrator vacillating between well-being and depletion, until the final showdown with Mrs. Briss. Then, as she closes the door in his face, telling him unequivocally that his scrutiny must stop, he realizes his poverty. "I, losing [her

untouched splendour] while, as it were, we closed a certain advantage
I should never recover, had at no moment since the day before made
so poor a figure on my own ground" (160). "My personal privilege, on
the basis of full consciousness, had become, on the spot . . . more than
questionable. . . . What did this alarm imply but the complete reversal
of my estimate of the value of perception?" (131).

The narrator ends his weekend with neither knowledge nor personal enrichment. His curiosity has alienated everyone and he is
thrown back on his own resources. Unlike Hawthorne's "steeplejack,"
who ended his story when he could no longer observe the panorama of
life, Jame's protagonist begins to embroider on the thin film of half-
truths at his disposal. What he has seen, meager though it is, is enough
to initiate a wealth of drawn-out fantasies, so that his "story" takes
form over and above the germ of observed facts, and indeed seems to
flourish on poor soil:

> A part of the amusement they yielded came, I daresay, from my exag-
> gerating them—grouping them into a larger mystery that the facts,
> as observed, yet warranted; but that is the common fault of minds for
> which the vision of life is an obsession. The obsession pays, if one
> will, but to pay it has to borrow (30).

What is clear to us, though it is not to the narrator, is the futility of
his initial goals—to acquire intimate knowledge without subjective
involvement, to support a theory about that which is inaccessible to
normal social exchange without imaginative embellishment. In fact,
James suggests that ignorance is the inevitable result of the narrator's
method and that invention is, in turn, a necessary consequence of ignorance: what the narrator does not understand from experience he cannot gain from observation and so he must fabricate.

Because of this concern with the parameters of objective observation, *The Sacred Fount* can be considered James's oblique answer to the
rules of composition advocated by the French naturalists: that facts
should speak for themselves, that the author should be completely unin-
trusive. Zola had said that one should stand before one's subject like a
photographer; but in James's opinion, such passive, affectless seeing was
impossible. No matter how detailed or precise, observation was always
inflected and limited; the eye alone could not take possession of the
meaning of a scene. With regard to the issues of observation, *The Sacred
Fount* is basically about the epistemological limits of the outside view;
it is James's dramatization of the consequences of disengagement, his
fictional illustration of vision becoming the measure of its own
estrangement. In these matters the camera provided him with a meta-

phor for identifying the separation of the seer from his own visual field, with all of the liabilities, distress, and need for imaginative reconstruction that alienation required.

V

The stridency of *The Sacred Fount* is that of a case taken to extremes. One could never claim that James endorsed his protagonist's "vampire theory," but the very existence of the theme shows James's sensitivity to the dynamic character of observation and his judgment that visual transactions were ones in which the artist stood to gain a great deal at someone else's expense. But if the book presents an extreme case, it nonetheless presents a common one, for James returned to the subject many times, turning it, returning it, examining its variations, even in novels of very dissimilar tenor.

The Princess Casamassima (1886) is such a book. Superficially it is a study in thematic contrasts: where *The Sacred Fount* takes the leisured upper class of Victorian England as its subject, the later text draws on the lives of London's working-class poor in 1881, a year of incipient social turmoil in the history of English labor relations. But in one respect the books share an imaginative substructure, for they are both about the effects of patronizing curiosity. The distinction between them lies in the later novel's expanded realm of action, for the prying that had been one man's private and idiosyncratic preoccupation in *The Sacred Fount* acquires political ramification in *The Princess Casamassima*. It is for this reason that it is a pivotal text in constructing a genealogy of the forms of aesthetic approach in American fiction. The voyeurism of the *Fount's* idle aesthete becomes in James's London novel the voyeurism of a sociological meddler, the observation of one class by another, and thus a more thorough and convincing demonstration of the power relationship behind supposedly objective interest. As long as James believed his preparatory spying to be psychological, it was open to dismissal: what is private can be construed as idiosyncratic and harmless. With this book, James moves the critical issues of the observational stance advocated by realist artists into the public sphere, where it is no longer possible to ignore the implications of asymmetrical social relationships. The genius of James's novel comes from his illustration of the reiterative rather than the confrontational structures of art and the social world that it seems, more overtly, to criticize. By this time in his career James understood the delusion involved in pitting the artist against a political power structure, in thinking which took economic poverty as its subject to be without its own kind of domination. For him, the very act of creation, predicated as it was on observation—or, as Foucault would say, "the gaze"—duplicated rather than opposed the mechanisms of power that subjugated and controlled the poor. He understood himself to be engaged in a displaced mode of oppression where vision or perception itself became a new social currency, a sign of asymmetrical exchange. Aesthetics, as he practiced it, conformed to a model which was continuous with a larger area of political and social transaction. To witness was not to be an impartial observer. To witness was, in this book, to control.

If we examine the text, we discover a familiar mode of emplotment: a young bookbinder, who is trying to understand the ramifications of a very confused personal heritage, is taken up and then dropped by an idle society woman. The book focuses clearly on the young man, but part of his interest is provided by the princess's interest in him, for he is one of the few working-class people to whom she can speak. As he struggles to understand his own class allegiances and to participate in the subversive politics of the London underground, he arrives at a complex and deeply informed sense of the relation between beauty and wealth and of the destruction revolution implies. In simple, reductionist terms, Hyacinth Robinson expresses James's sense of the irreconcilable division between art and social responsibility. He is the one person in the book who gives to each of these aspects of life a complicated and vexed attention, and who is unable to act without a clear allegiance to one or the other. Throughout the book he is called "the little bookbinder"; he is condescended to and used by those who cannot understand the seriousness of his personal dilemma, but James endorses his character's complex sensibility and shows his appreciation of art to be all the more extraordinary because it is not fostered by a privileged class background and training.

Robinson dies from his irresolution: unable to murder a government official in the service of revolution, he is also unable to bear the class betrayal that his seeming cowardice implies, and so he takes his own life. For James, Robinson's tragedy lies in his inability to simplify and sacrifice arbitrarily one aspect of life for another equally valuable social goal. But he fills his text with those whose sensibilities are not so complex and who pull the young man one way or another according to their own more singular ambitions. For all her loveliness and charisma, the Princess Casamassima is one of these philistines—she is someone whose good intentions, coupled with a dearth of self-reflection, become actively harmful. And it is she, more than Robinson and more even than the conspiring revolutionists, who is the principle observer in the book: "[T]hen restlessly, eagerly . . . she arrived at the point that she wanted to know the *people,* and know them intimately— the toilers and strugglers and sufferers—because she was convinced that they were the most interesting portion of society, and at the question, 'what could really be in worse taste than for me to carry into such an undertaking a pretention of greater delicacy and finer manners? If I must do that,' she continued, 'it's simpler to leave them alone. But I can't leave them alone; they press on me, they haunt me, they fascinate me. There it is—after all it's very simple: I want to know them and I want you to help me.'"[42]

For James, many of the book's most compelling problems are discussed through the princess's interest in the lower classes. Born in comfort and married to a wealthy man, ostensibly she falls into a category familiar to the 1800s, that of the settlement house worker—someone like Octavia Hill in London, who concerned herself with the housing of the poor, or like the young woman described so aptly by the American, Jane Addams: a well-brought-up girl who is troubled by her own uselessness and "restless about being shut off from the common labor by which [she lives] and which is a great source of moral and physical health."[43] But where Addams was significantly aware of the subjective motivations for helping the poor and of the reciprocal benefits of such work, James's protagonist lives without commensurate self-knowledge. In fact, much of the book is dedicated to the gradual revelation of the personal and idiosyncratic nature of motives that are ostensibly selfless. However short-sighted we may think Jane Addams's approach to the problem of urban poverty, it was at the least an earnestly espoused and reflective position. By contrast, the activities of James's princess assume a sinister cast, and part of their dubiousness arises because she assumes a spectator's role in relation to others: Robinson is never taken seriously; he is not befriended to be a friend, but to be a performer or a specimen of poverty. It is not surprising, then, that she is disappointed when he belies her stereotype of deprivation. "'The only objection to you individually is that you've nothing of the people about you—today not even the dress. . . .' Her eyes wandered over him from head to foot and their recognitions made him ashamed. 'I wish you had come in the clothes you wear at your work'" (I,292). In the same way that the narrator of *The Sacred Fount* engaged in social interaction in order to elicit proof of a theory, the princess submerges herself in the fates of London's poor in order to relieve the boredom of her own more socially constricted circumstances. "You'll not persuade me either that among the people I speak of characters and passions and motives are not more natural, more complete, more *naïfs.* The upper classes are so deadly *ban-*

als. . . . Or if you knew what I've been through you'd allow that intelligent mechanics . . . would be a pleasant change" (I,201). She is playing at an involvement that has no hereditary roots and no basis in present reality. As the other characters come to realize before she does, no amount of acquired knowledge can make up for her differences in status. However sympathetic she may be, she, like Lady Aurora, will always be a privileged outsider:

> The only fault she had to find with these latter [the Muniments] was that they were not poor enough—not sufficiently exposed to dangers and privations against which she could step in. . . . He didn't mind, with the poor, going into questions of their state . . . but he saw that in discussing them with the rich the interest must inevitably be less: the rich couldn't consider poverty in the light of experience. Their mistakes and illusions, their thinking they had got hold of the sensations of want and dirt when they hadn't at all, would always be more or less irritating (I,316).

James allows his character to resist this view, to believe that empathy can create a bond of experience between dissimilar people. "She had been humiliated, outraged, tortured; she considered that she too was one of the numerous class who could be put on a tolerable footing only by a revolution" (I,293). But she continues to act, perhaps unwittingly, in ways that prohibit making that bond complete. She pumps Robinson: "[S]he didn't wish vague phrases, protestations or compliments; she wanted the realities of his life, the smallest, the 'dearest,' the most personal details" (II,127). But eventually the bookbinder recognizes the solipsism behind her overt concern, the personal solution sought through supposed disinterest: "For the moment, nonetheless, her discoveries in this line diverted her as all discoveries did, and she pretended to be sounding in a scientific spirit—that of the social philosopher, the student and critic of manners—the depths of the British Philistia" (II,39). To Robinson she admits, "I'm determined to keep hold of you simply for what you can show me," and he, amid all his desire to think well of her, comes to realize that "she was in earnest for herself, not for him" (II,127). The entertainment afforded by her dabbling in poverty is revealed eventually by her desire to *pay* for picturesque sights: "On the Sundays when she had gone with him into the darkest places, the most fetid holes in London, she had always taken money with her in considerable quantities and had always left it behind. *She said very naturally that one couldn't go and stare at people for an impression without paying them*" (II,260) (my italics).

In time we come to recognize the pattern in these fictional events and to understand that James has established an analogy between the nature of the social order in *The Princess Casamassima* and the individual psyche in *The Sacred Fount*. In both novels, appearances are foils which require penetration.

When she interests herself in the budding proletarian movement, the princess speaks of the revolution as a "subterranean crusade," or as something brooding "beneath the surface." She assumes the social order as a whole to have an agitated interior which is obscured by social forms, and her interest in Robinson stems from her need to have a guide to a world to which she would normally be denied access. That internal social distress, the hidden agitation of the working class, is described in terms that are unnervingly like those that define May Server's private grief in *The Sacred Fount*, where the narrator muses on "conditions so highly organized that under their rule her small lonely fight with disintegration could go on without the betrayal of a gasp or a shriek" (167).

By setting up this parallel between collective social and private psychological behavior, James establishes an essential connection and moves the seeing that is the central activity of the realist novel into a fuller political dimension.[44] Psychological probing and social probing come to reflect each other; the interest of the secure world in the nether world is shown to be similar to the curious individual's interest in the private lives of others, and both serve James as oblique references to his own creative situation. The "seeing" he does as he walks the London streets, quarrying the underworld for material, is analogous to the "seeing" enacted by the photographers of his time who searched so diligently for the picturesque: "Face to face with the idea of Hyacinth's subterraneous politics and occult affiliations, I recollect perfectly feeling, in short, that I might well be ashamed if, with my advantages— and there wasn't a street, a corner, an hour of London that was not an advantage—I shouldn't be able to piece together a proper semblence of those things, as indeed a proper semblence of all the odd parts of his life."[45]

In all cases we can see what may be called an aestheticization of experience—a distancing which allows the picturesque nature of some aspect of life to be located and framed. In each case, people become objects of contemplation; and in the case of the Princess Casamassima, the poor are easily targeted because they are undefended and thus available for the princess's scrutiny. Hyacinth Robinson may long to see the lives of the rich, but he cannot take a look; he must wait until doors are voluntarily opened to him. Though the novelist consistently posits life as theater and proceeds by constituting scenes of mutual entertainment, only the wealthy and powerful can choose the stage on which they will stand, only they can choose their entrances and exits. The poor, the laboring classes, the inhabitants of the streets and pubs stand always vulnerable to the gaze of curious interlopers.

One could not claim that this aesthetic tendency illustrates the camera's influence per se, but it is important to note that it is a model of relating to the world, shared by photographers and writers alike. In James's case, for example, one can easily see that passivity coupled with an eye for the potential utility of the scene was a posture shared by Coburn. With all of Coburn's sensitivity to the mutual influence of photographer and subject, he unthinkingly divorced vision from responsibility. Though he never engaged in documentary work, Coburn was fascinated by poverty and routinely searched for scenes that might normally be considered ugly, seeing in the arrangement of buildings or the accidental display of light on a squalid tenement the basis for a beautiful composition. "I have photographed in many industrial cities . . . always with enthusiasm and interest. Photography makes one conscious of beauty everywhere, even in . . . what is often considered commonplace or ugly. Yet nothing is really 'ordinary,' for every fragment of the world is crowned with wonder and mystery."[46] One can see in all of these instances—in James's two stories, in Coburn's "ugly" photographs—a tendency toward disengaged observation, a propensity to dissociate vision from action or at least from responsible interaction with the subjects of the scene. As long as beauty or composition was Coburn's goal, his concern for poverty-stricken streets extended only to the edge of his view-finder.

It is in this procedural tendency that James showed the greatest affinity with the camera. Though he struggled to make formal distinctions between his fictional creations and the work of photographers—distinctions that were, as I have shown, much less pronounced than he would have allowed—probably he would not have balked at admitting the community of ethical problems that bound photographers and fiction writers together.

Through the use of characters within his fiction to reflect or reenact his own creative dilemma as a textmaker, James repeatedly demonstrated the culpability he felt to be inherent in using sight as a creative resource. The princess watching Hyacinth Robinson is, after all, remarkably like the photographer in Sadakichi Hartmann's brief story: for both of them, the "direct impression" is gained by a profound personal neglect. We recall Hartmann's pointed exchange: "'But is it not wonderful that there is a whole world around us to look at for years and years? And yet we are never aware of it, we never see it until some happy moment suddenly reveals it to us. And I owe it all to you.' 'By ignoring me,' she said, reproachfully." And we remember that the price of the Princess Casamassima's "ignoring" Hyacinth Robinson, her failure to intervene or care about him as more than an interesting case of poverty, is his death. That James again takes the problem of disinterested vision to its most extreme form is perhaps a measure of his own discomfort. That he faced the issue so clearly and without excuse is also a measure of his complex sensitivity, an admission of an unresolvable dilemma, and a covert desire for expiation. The fictional reenactment of damaging disinterest tells us emphatically that the "expense of vision" was what Henry James could neither avoid, nor, to his credit, refrain from confessing.

VI

The lasting importance of Henry James's first proletarian novel comes from its reenactment of the power relations implicit in the very act of viewing the world. In his preface to *The Princess Casamassima,* James remembers himself as a night prowler, a walker of the London streets, a person of passive but receptive sensibility: "I walked a great deal ... and as to do this was to receive many impressions, so the impressions worked and sought an issue, so the book after a time was born" (I,7). He would have us understand the artist to be a ghostly presence, no more intrusive than is the pavement upon which he walks in search of material. "I recall pulling no wires, knocking at no closed doors, applying for no 'authentic' information," he said, but he also recalled that "on the other hand the practice of never missing an opportunity to add a drop, however small, to the bucket of my impressions.... To haunt the great city and by this habit to penetrate it ... that was to pull wires, that was to open doors" (I,19-20).

What would have happened to James had he tried more actively to open doors to the London underground and to the militant socialist or anarchist movements of which he imagined his protagonist to be a part? A reporter for the *Evening Standard* of the time who wrote a piece entitled "The Haunts of the East End Anarchist" found out when he tried to go to the "Workers' Friend" in Berner Street: "If you are unknown to the speaker, you will be told that no business is done there. If the questioner recognizes you, or you come with a friend, a string of arrangements will open the side door."[47] Henry James would have been denied access, and in that refusal would lie the structural truth of his entire position: those whose secrets he wanted to share would have known the deception of his neutrality, would have recognized the power of knowledge itself, and would have resisted his approach. The anarchists would have immediately made the connection that James so poignantly but yet too pointedly denies: that vision and supervision are experienced identically. To view the social scene is to have the potential to dominate it, to subject it to the will, to reinscribe a pattern or strategy of subjugation. It is precisely in address to this issue that *The Princess Casamassima* becomes crucial to a new understanding of literature predicated on investigation. James's fictional princess embodies a creative posture for James, but in addition, she exemplifies a structure of knowledge that the late nineteenth century was in the midst of inventing to deal more generally with the distress and social unrest of the urban poor of London. No matter what disclaimers James may make for himself in the preface to *The Princess Casamassima,* the "seeing" that constitutes the major activity of this text is a displaced codification of the structures of observation and control that characterized the nascent social science industry of the time; and the text's brilliance, predicated as it is on a certain understanding of the dynamics of watching and of finding the ugly to be picturesque, lies in its demonstration of the strategies of manipulation and concealment that underlie and unite all these discourses—scientific, social scientific, and aesthetic. In all cases, the crucial issue was the cognizance taken, and the effects of that gaze. What James and Coburn let us see is the fallacy of thinking the photograph or the novel to be autotelic and self-made, without human intervention. They explode the ideology of neutrality which was at the heart of the early social science investigations into poverty and illustrate the political status of supposedly value-free methods.

Let us look for a moment at the social science literature surrounding James's novel. It is impossible to discover which, if any, of these books, essays, and newspaper articles James read—he consistently maintained that his work was predicated on the "direct impression"—yet his text unmistakably belongs to a genre. In saying this, I do not mean genre in Northrup Frye's sense of a predictable emplotment, but rather a group of books characterized by a common thematic concern and by a common formal strategy in the face of it. Subject matter per se is not of interest here, for the theoretical issues that are raised by the realistic fiction of this period seem to continue only peripherally the debate about the "low" or "base" or "animalistic" focus of such art and to confront more centrally the issue of vision as a strategy of power, an investigative posture, an asymmetrical relation between an observer and a subject.

The Princess Casamassima, however unique in the James canon, was another rendering of a peculiar but, at the time, common kind of urban spy literature. In these books, the protagonist pitted him or herself against the unknown and presumably horrible circumstances of poverty, and survived by playing raw intelligence against the rot and degradation of the city. James himself speaks of there being "London mysteries (dense categories of dark arcana) for every spectator" (I,8) and then comments, rather curiously, that it is a "state of weakness to be without experience of the meaner conditions, the lower manners and types, the general sordid struggle, the weight of the burden of labour, the ignorance, the misery and the vice" (I,8). To know poverty is to acquire a kind of status. In two brilliant essays on the literature of the city, Alan Trachtenberg and Mark Seltzer have discussed this "spy mania" and the darkness of the city as a challenge to middle-class intelligence.[48] In America, the young Stephen Crane disguised himself as a bowery bum in order to write "An Experiment in Misery." How, he asked, did such derelicts live each day; how did they survive without money; what were their thoughts as they met an ill-chosen fate? In London, the most famous of these adventures was undertaken by another American, Jack London, who reiterated the experiment with more elaborate preparation and cunning. In *The People of the Abyss,* he explains:

> The experiences related in this volume fell to me in the summer of 1902. I went down into the underworld of London with an attitude of mind which I may best liken to that of an explorer. I was open to be convinced by the evidence of my eyes rather than by the teachings of those who had seen and gone before.[49]

His text is a curious study of the literal and psychological defenses against the very experience he courts: the "anarchists, fanatics and madmen" of the London underworld attracted and repelled him, took shape primarily in response to him, and were, finally, the subsidiary subject. The drama of the book, as he well knew, was self-dramatization: London's essential exploration was of himself as he fell purposely from grace, assuming the ragged clothes of the poor so that he could assume their companionship and experience, so that the reaches of art might be extended by the reaches of his own exotic adventures in workhouses and pubs.

To some extent, he simply reported what he discovered of their lives, adding to the literature of case histories. Of a Kensington family that had moved to London, he said:

> They fell into the hands of those who sweat the last drop out of man and woman and child, for wages which are the food only of

despair. . . . The drink demon seized upon them. Of course there was
a public house at both ends of the court. There they fled . . . for shelter, and warmth, and society, and forgetfulness. . . . And in a few
months the father was in prison, the wife dying, the son a criminal,
and the daughters on the street. Multiply this by half-a-million, and
you will be beneath the truth.[50]

On occasion, he spoke for them: "They feel, themselves, that the forces
of society tend to hurl them out of existence." But his most customary
approach was to mingle this reportage with commentary the covert
function of which was to distinguish the speaker from the subject of
his own observation, to assert intelligence and analytic power as assets
as tangible as the book contract and bank account which held him personally above "the abyss." London's preparations for this venture were
as elaborate as those of any political mole: he assured his identity with
a detective so that, even in ragged disguise, he would never be mistaken
by the police for a desperate and destitute man. He arranged a safe
house, a "port of refuge" into which he could run "to assure myself
that good clothes and cleanliness still existed. Also in such a port I could
receive my mail, work up my notes, and sally forth occasionally in
changed garb to civilization."[51] He also carried hidden money, and it
was this circumstance that most clearly showed him the connection
between class status and confidence, between camraderie and trust: he
used it only to relieve the hunger of two other men who, like himself,
had been refused entrance to the Poplar Street workhouse, but its existence changed the entire balance of their relationship. "I had to explain
to them that I was merely an investigator, a social student. . . . And at
once they shut up like clams, I was not of their kind; my speech had
changed, the tones of my voice were different. In short, I was a superior,
and they were superbly class conscious. . . ."[52] London assumes here that
it is his own superiority in status that stifles the conversation, but
though he is correct in this, he does not pursue the ramifications of his
own logic. Neither does he make the connection between this experience and the theory he eventually offers to explain the collective misery of these East End men. Jack London's belief was always in the
blindness of fate and in the unkind and overwhelming effects of the
environment. In both his fiction and this strange study, he represented
the poor as helpless in the face of raw, natural force, but he did not see
that his own presence among them had the same affect as that unname-
able force. His surveillance, once recognized, was itself powerful
enough to engender the silence of companions who knew it to be their
only defense against his mystifying advantages. For what reason could

he be among them? London did not see that the very act of reporting caused feelings of helplessness which he later came to think of as responses to disembodied and thus unchallengeable fate. That is, he unwittingly embodied the very structural situation he thought to exist outside his perception of it.

Mrs. Cecil Chesterton and, oddly, William Dean Howells, were two other writers who experimented in these ways. Howells's street walking resulted, as it had in New York, in a pedestrian and audaciously uninformed account of the city, called *London Films*. Though he apologized for his own banality through the metaphor of this title (he was carrying "a mental kodak with him" which could "represent [only] the surfaces of things"), his contentment with these surfaces, his satisfied opacity, was extraordinary. His handling of the "lowly and the lowlier" was left with another apology: "The friend who had invited me to this spectacle [the Jewish market] felt its inadequacy so keenly . . . that he questioned the policemen for some very squalid or depraved purlieu that he might show me, for we were in the very heart of Whitechapel."[53]

Mrs. Chesterton, a well-heeled and well-intentioned woman, wrote a more reflective book called *In Darkest London* in which she claimed to have "walked with other souls in Pain." In essence, she repeated Jack London's experiment from a woman's perspective: "I wanted to find out how the woman without a home, without a reference, without money, friends or a decent wardrobe, supports life. . . . I left my home one evening in February . . . I would arrive in London with nothing but my personality between me and starvation."[54] I quote from Howells's and Chesterton's books primarily to show the reiterative nature of these ventures, the fascination of the slums, the perceived need for the concealment of origins in order to explore them, and the idiosyncratic or unsystematic nature of each investigation. Yet Mrs. Chesterton's book is obviously an oddly twisted roman-à-clef, a book whose very title refers back to one of the earliest and most famous sociologies of the urban poor, General [William] Booth's *In Darkest England and the Way Out*.[55] It takes its place among what Gareth Stedman-Jones has called "the growing literature on 'outcast London,'" a professional literature of case studies, multiplied into statistics, undertaken in the face of growing social crisis and uncertainty about the mood of the working classes.[56] In 1885, Samuel Smith wrote: "I am deeply convinced that the time is approaching when this seething mass of human misery will shake the social fabric, unless we grapple more earnestly with it than we have yet done. . . . The proletariat may strangle us unless we teach

it the same virtues which have elevated the other classes of society."[57]
His warning was both an outgrowth of, and an admonition to, super-
vision, a response to vaguely located anxiety, and to an emerging crisis
in the 1880s, which, again according to Stedman-Jones, was composed
of four elements: "A severe cyclical depression as the culmination of
six or seven years of indifferent trade; the structural decline of certain
of the older central industries; the chronic shortage of working-class
housing in the inner industrial perimeter; and the emergence of socialism and various forms of 'collectivism' as a challenge to traditional
liberal ideology."[58]

My purpose here is not to offer a full analysis of the historical
situation that surrounded Henry James when he wrote *The Princess Casamassima* in the early 1880s, but to show that his text is part of a con-
tinuum of inquiries into "misery, vice and crime among the poor" and
that their common substructure arises out of a shared mode of access
or approach to the subject. When placed against this background,
James's princess becomes an idiosyncratic, almost parodic, version of
Mrs. Chesterton, Jack London, and ultimately William Booth, Henry
Mayhew, George R. Sims, and Adolphe Smith,[59] all dipping down into
the London underground, disguising their original identities because
their reasons for looking around have more to do with their own welfare than with actual concern for their subjects. Samuel Smith said quite
openly that his real fear was that misery might "shake the social fabric." If we look for a collective motive for spectatorship, it was not
disinterest, but containment. Vision served the interests of policy;
reform measures—and we read in the newspapers of these years column
after column describing the Employers' Liability Act, or the need to
address the "smoke" question or the "noxious gas" question or the "tenement house" question—were undertaken in the interests of circum-
scribing and subduing unrest. In his own text, James reinscribes all of
these relationships in order to show that the imagination can have its
own structures of inquiry. The Princess Casamassima, reiterating the
postures of her social science precursors, worries that the victims of her
scrutiny will regard themselves as "curious animal[s]." She eats them;
she preys on them; she takes them into herself for herself. But Hyacinth
Robinson eventually figures out that he is, despite the princess's disclaimers, nothing more than an oddity. He comes to see the falsity of
her motives. If his suicide can be interpreted as an inability to enact the
violence he has pledged himself to, it can also be seen as a testament of
power that is insufficient to consciousness. Robinson's knowledge of his
subjugation is inadequate to the task of rebellion. And in his struggle

and defeat we can read the resentment of the working classes, their awareness of surveillance and their admission of helplessness in the face of it. In this respect, *The Princess Casamassima* finally is about the failure of imaginative discourse to discover a viable mode of social reciprocity. It shows us both a privileged mode of discourse and a discredited mode of discourse, a subjugating will and a failed resistance to it. In his attention to the frustration of the unempowered, James anticipates those novelists who eventually would give voice to mute and suffering people, raising them to the level of self-reflection, and granting them a status commensurate to their own evaluation of life's circumstances. His book points to the achievements of a later generation of writers who, like Theodore Dreiser, would write from the perspective of damage in address to those who would prefer to contain them within the safety of an imaginative frame. "Curious animal" that he was, Dreiser refused to be either silent or picturesque, protesting his peripheral status with a vehemence that would have astounded Henry James, even though James had struggled to represent the very conflicts that gave rise to this resistance.

Alvin Langdon Coburn.
Faubourg St. Germain.
Paris, 1906
(IMP/GEH)

Alvin Langdon Coburn.
Portland Place.
London, 1906 *(IMP/GEH)*

Alvin Langdon Coburn. The Venetian Palace. Venice, 1906 *(IMP/GEH)*

Alfred Stieglitz. The Steerage. 1907 *(IMP/GEH)*

3

Theodore Dreiser,
Alfred Stieglitz, and Jacob Riis
Envisioning "The Other Half"

> He knows because his whole experience is one of
> being excluded from the exercise of power in his soci-
> ety and he realizes from the compulsion he now feels,
> that art, too, has a kind of power. —John Berger

I

WHEN Henry James returned to the United States in the
summer of 1904, he returned to a world that seemed
unpredictable and disturbing to him. "This sense of dispossession ... haunted me," he admitted, and his travels acquired
urgency as he tried to reconcile memories of his childhood home with
scenes of contemporary American life. "The manners, the manners:
where and what do they have to tell?" he asked. In no other place was
he more curious and anxious to read the signs of American cultural life
than he was in New York City; and in no other place was he at such
a loss to understand what he called "the social mystery, the lurking
human secret."[1]

His motivation was once again the obsession of a "story seeker."
Like his earlier protagonist, the Princess Casamassima, he explored the
streets of the city, sensing that the sheer density of human life contained a profound, if enigmatic, message. Without a guide to the underworld, he remained always at the edge of that enigma, baffled by the
constant activity of the streets, and finally thrown back upon his own

93

estrangement. For Henry James, "the problem of the alien" was as much the problem of his own alienation as it was a description of teeming immigrant life.

What struck him most deeply was the city's energy. "It is the power of the most extravagant of cities," he claimed, "rejoicing as with the voice of the morning, in its might, its fortune, its unsurpassable condition" (74). But his hyperbole faltered when he tried to penetrate the turmoil, to understand New York's place among other cities and to grasp the human implications of its size. When it came to specific qualities, James mourned the city's ugliness, its mutability, its destruction of tradition: skyscrapers towering over churches told him that commerce took precedence over things of the spirit; buildings under construction announced the city's constant search for the new; the whole scene was ugly, but it did not cohere, as did older European cities, into something picturesque. Like John Ruskin in *The Stones of Venice,* James repeatedly extricated signs of individual human effort from this overwhelming and abrasive energy. He delighted in the small church in the midst of towers, the detailed bit of architecture lost in a sea of concrete and granite, the particularly arresting face. He thought he had an "almost inexpressibly intimate" relation to New York.

Yet it was a tenuous relationship, built on an appreciation of what was past or passing. The America that confronted James in 1904 was intent on growth and change. Where Jacob Riis, members of various tenement house commissions, and, indeed, most of those involved in the incipient social reform movement in America spent years trying to tear down the nastiest buildings of the lower East Side, to bring sunlight and fresh air to people held captive by filth, James complained that Americans had no reverence for the buildings and monuments of the past: "Let sentiment and sincerity once take root; let any tenderness of association once accumulate or any 'love of the old' once pass unsnubbed, what would become of us?" (112). He saw American workers in a similar light, for they disrupted what he thought of as traditional ways of relating to others. He was at a loss when the reciprocal obligations of men were not clearly defined by inherited mannerism. "There is no claim to brotherhood with aliens," James insisted, although he made forays into the Bowery to search for connections and presumably to discover why "the social question quite fills the air in New York" (120, 114).

On one of his first visits, he discovered a surprisingly high standard of living. "My friend was quartered, for the interest of the thing . . . in a 'tenement house.' . . . On my asking to what latent vice it owed its stigma, I was asked in return if it didn't sufficiently pay for its name by harbouring some five and twenty families [but] conditions [were] so little sordid, so highly evolved" (135). Since it contained a white marble staircase, the building told James that the furor over substandard housing was inflated, that disembarking foreigners moved with grace into a waiting community of their own ethnic background and that opportunity and a "diffused sense of material ease" characterized most people's lives. His private terror was in the imagined future when these people would reduce everyone else to a "common denominator," when the "vulgar assault of the street" would accomplish its work fully. The writer who made so much of the lives of the great houses and villas of Europe reduced the importance of New York's poverty-ridden tenements by understanding them to be little more than picturesque sights offered to a refined tourist. Instead of providing balance or an added dimension to the issues before him, James elevated idiosyncrasy into generalization, letting high culture speak for and obscure culture in any other form. It was sufficient to "look-in" as preparation for his text: "[In the Bowery] the question, for this second visit, was of a 'look-in' with two or three friends, at three or four of the most characteristic evening resorts of the dwellers of the East Side. It was definitely not, the question, of any gaping view of the policed underworld" (200). He did not see that poverty could have a voice of its own; for him there was no culture of distress.

However narrow Henry James's conclusions about New York's immigrant population may have been, he was nonetheless correct in granting "the social question" proper weight. "The problem of the alien" was visible everywhere; it was the collective dilemma of New York City as it entered the twentieth century. Very few people found it possible to respond to poverty solely as the Princess Casamassima had responded to Hyacinth Robinson in fictional London or as James responded in fact to New York's Bowery, for urban blight and those damaged by it occasioned fear as well as aesthetic interest and curiosity. In *The Dangerous Classes of New York* (1872), Charles Loring Brace, evangelist and reformer, had predicted an "explosion from this class which might leave the city in ashes and blood." "It has been common, since the recent terrible Communistic outbreak in Paris, to assume that France alone is exposed to such horrors; but in the judgment of one who has been familiar with our 'dangerous classes' for twenty years, there are just the same explosive social elements beneath the surface of New York as of Paris."[2]

Where James had seen in New York a social and architectural drabness that threatened to reduce a formerly treasured liveliness of scene, Brace recognized a threat of a much more fundamental order. For both,

however, the consequent problem was the same: how to read and interpret several orders of evidence about the streets, the interiors of houses,
and the nature of human life harboured in these places. For one, the
need was occasioned by art; for the other, scrutiny occurred in the
interest of social control and the protection of property. Brace knew
very clearly that he worked "in the interests of public order, of liberty,
of property." What is interesting about both men is their tendency to
polarize the social world into a place sharply divided into "we" and
"they" and their common assumption that universal validity adhered
in their own descriptions and interpretations of what they saw. But
neither man spoke impartially. Brace was an advocate, not a scientist,
and James's status as artist-observer did not exempt him from bias. If
his perspective had any collective status, it was as a representative of an
established social group the security and values of which were threatened by the "barbarism" of the tenements. Though James spoke deci-
sively about social conditions, he also spoke in a voice that he suspected
would become obsolete. As his friend and colleague, Henry Adams,
observed, "[H]e did not presume to be unique; he belonged to the type."[3]

II

Theodore Dreiser entered upon this scene with an intensity and hunger
equal to Henry James's. He, too, understood that the life of the city
would provide one of his major themes, but he had no need to "look-
in" on poverty. Even when he was newly arrived from the Midwest,
Dreiser recognized the deprivation that lurked behind New York's tenement walls. Writing "The Prophet" in 1896, he said:

> These endless streets which only present their fascinating surface are
> the living semblance of the hands and hearts that lie unseen within
> them. They are the gay covering which conceals the sorrow and want
> and ceaseless toil upon which all this is built. They hide the hands
> and hearts, the groups of ill-clad workers, the chambers stifling with
> the fumes of midnight oil consumed over ceaseless tasks, the pallets
> of the poor and sick, the bare tables of the hopeless slaves who work
> for bread. Endless are these rows of shadowy chambers, countless the
> miseries which these great walls hide.[4]

Hawthorne had stood mutely before the barrier of walls; James had
haunted the scenes of the tenement world, hoping for the key to its
mystery, but brick hid nothing from the young Dreiser. The son of an

itinerant worker from Terre Haute, Indiana, he had been born in poverty and wanted, if anything, to distance himself from it. To the great debate about "the social question" in New York, Dreiser brought a complex set of desires and the need to find a mode of writing adequate to the expression of those desires. One might say that he did not initially know from which "half" of the tenement wall to write: should he express the world as he experienced it among the dispossessed or should he write as if he were already established in the cultural community which harbored writers like Henry James? For Dreiser "the social question" was directly and urgently a question of how to maintain a kind of dual consciousness in the face of the complexities posed by his need to name and identify himself on both the personal and cultural levels. On one hand, he could see the virtue of establishing a distant perspective, like that of Alfred Stieglitz, who thought the poor to be picturesque and who made immigrant life into a private emblem of the vitality, courage, and energy he missed in the bourgeois activities of his family and peers. Speaking of the achievements of pictorial photography in 1899, Stieglitz had insisted that he presented "metropolitan scenes, homely in themselves . . . in such a way as to impart to them a permanent value because of the poetic conception of the subject displayed in their rendering."[5] On the other hand, Dreiser could understand the work of someone like Jacob Riis who used the camera to demolish the "[b]arriers erected by society against its nether life." "I am not willing," Riis had said, "even to admit it to be an unqualified advantage that our New York tenements have less of the slum look than those of older cities. It helps to delay the recognition of their true character on the part of the well-meaning, but uninstructed, who are always in the majority."[6] He would have been appalled at Stieglitz, for in his eyes "homely" metropolitan scenes were never neutral and their "poetic" expropriation would not have been viewed as an apolitical act. To him, the picturesque nature of poverty presented by Stieglitz's camera would have simply reinscribed the ideology that made the poor seem heroic even in the midst of their degradation. Where Stieglitz spent his life dissociating photography from its utilitarian function, Riis used the camera to present evidence; where Stieglitz wanted to wrest his images from all political imperatives, to decontextualize the photograph—or at least firmly to recontextualize it among the fine arts—Riis embedded his images in a discourse about liberal reform and used vision in the interest of policy decisions.

Their photographs of Manhattan at the turn of the century were so distinctive that they could quite possibly have come from different cities. Yet Stieglitz's particular way of photographing immigrant life had as many interesting ramifications as his colleague's unhesitating alignment of the camera with the nascent social science movement. Both of them helped Dreiser to assess his own relation to a poverty that he recognized but despised and to find the representational strategies that could most authentically express the difficult and largely unprecedented literary perspective which later made him such a figure of controversy on the literary scene.

III

Dreiser met Stieglitz shortly after arriving in New York. In 1899 he was not yet known as an aspiring writer; he was, in fact, a year away from publishing his first novel, *Sister Carrie,* but he was to become one of the most famous and controversial realists of his generation. In one of his first ventures into journalism, Dreiser concluded that Stieglitz was a "master of photography," a "sincere" and "patient" artist whose work was characterized by "that selection of subject, that delicacy of treatment and that charm of situation and sentiment which all rare paintings have." Of the twelve photogravures reproduced in Stieglitz's new portfolio, "Picturesque New York," he singled out three for special attention: "A Rainy Day on Fifth Avenue," "Reflections-Night," and "The L in a Storm." They were all street scenes, taken under trying circumstances which had required Stieglitz's most consummate skill in printing. It was as if the rain, the night, and the snow had called forth a delicacy in the artist commensurate with their inherent darkness and discomfort.[7]

Several years later (1902), Dreiser was to amplify his views about Stieglitz and photography in a second article called "A Remarkable Art: The New Pictorial Photography." It was a chatty and informative piece, intended to explain how certain kinds of photographic effects were achieved with bichloride of gum, and it was remarkable in itself for its sympathetic championing of the principles of pictorial photography: photography was an art commensurate with pictures done in oil, watercolor, and pastel; the photographer's brushes were "shafts of light," and his subject was always chosen with painstaking care:

> He will look for pictorial possibilities in the landscape. The bare, hard outlines of a scene are nothing in themselves. It is the beauty which he sees; the individuality of his own impression which he wishes to convey to others, and the subject which he selects must, in a way, embody these. No two men see the same element of beauty.[8]

With an almost dogged humility, Dreiser reiterated Stieglitz's techniques and opinions, honing his own perspective to accord with "the
master's," using whatever meager influence his journalism might have
to buttress Stieglitz's cause. Like his contemporaries, Hart Crane and
Sherwood Anderson, Dreiser fell under Stieglitz's charismatic spell, and
like theirs, his allegiance was poignantly evoked by desire. Stieglitz
represented what he most longed to achieve for himself: crusader for
truth, armed against the constrictions of middle-class convention, the
embodiment of bohemia; Stieglitz nonetheless spoke with the assurance
of a fully successful man. To the hungry and unproved young writer,
Stieglitz's pronouncements about individualism, beauty, and mood held
the promise of his own future.

At the end of his article, Dreiser pictured the older artist in the
"spacious quarters" of the New York Camera Club and also against the
larger background of the city, "the panorama of roofs and spires and
jetting steampipes, and the narrow grottoes of streets in the depths of
which the turgid stream of humanity flowed noisily." "'If we could
but picture the mood!' said Mr. Stieglitz." It was an emblematic setting.
Securely housed, both by reputation and by material circumstances,
shielded by the window which became the figurative eye of a camera,
Stieglitz turned lower New York and the harbor into a "fogged negative." As Dreiser understood even at this early point in his career, Stieglitz knew how to see and at the same time how to protect himself
from what he saw. Such a vantage point turned the "turgid," "noisy"
spectacle of New York into something quieter and less disturbing than
it might have seemed to a man less sheltered from its threats.[9]

If the younger man sensed an affinity with the subject of his interview, it lay in their temperaments, their distaste for bourgeois convention, and in their common obsession with watching the city: it is not
an accident that Dreiser remembered Stieglitz staring out of a skyscraper window at the environment around the Brooklyn Bridge. Years
later when he wrote the story of his life in journalism, Dreiser remembered his newspaper assignments as the means to more personal goals
that also involved urban spectacles: New York had been preceded by
Chicago, St. Louis, and Pittsburgh, but the fascination had always been
similar:

> And now, spread before me for my survey and entertainment was the
> great city of St. Louis, and life itself as it was manifesting itself to me
> through this city. This was the most important and interesting thing
> to me, not my new position. . . . The city came first in my imagina-
> tion and desires.[10]

Born in obscurity, destined for a life of continued small-town marginality, Dreiser's chosen profession became a way into an unknown
but more knowing world. At the beginning of his career, Dreiser knew
almost nothing of the practical world, and he seemed to equate social
knowledge gropingly and half-articulately with control of his own destiny. To know the news, to be on the scene, to watch the domestic
dramas of the city, its fires and accidents, was initially as close as
Dreiser got to understanding the origins and genesis of the American
social structure. But intuitively he aligned the desire to know with a
will to power, seeing a continuity between art and power rather than
a displacement or opposition of the political and the aesthetic.
Although by mid-career he had radically shifted his evaluation of the
power associated with observation, he had begun his professional life
with rather marked views about the gains to be had from the role of
observer. Like Hawthorne and James, he was driven to his position by
temperament and by ignorance, but unlike them, his motivation had
not been that of a voyeur or an onlooker in need of vicarious experience. If anything, Dreiser's drive to observe came from an obverse need
to find a comfortable distance from that which threatened to engulf
him because of his too great vulnerability or because of his unwieldy
sympathy with the downtrodden. When he first arrived in New York
in the winter of 1894–95, he spoke from their vantage point. At the
suggestion of his brother, Paul Dresser, Dreiser walked the streets,
acquainted himself with different districts of the city, and finally
paused at City Hall Park.

> About me on the benches of the park was, even in this gray, chill
> December weather, that large company of bums, loafers, tramps,
> idlers, the flotsam and jetsam of the great city's whirl and strife to be
> seen there today. I presume I looked at them and then considered
> myself and these great offices, and it was then that the idea of Hurst-
> wood was born. The city seemed so huge and cruel.[11]

Hurstwood would become one of Dreiser's most notable fictional characters whose initially affluent Chicago life ended in the Bowery.
Dreiser uses this memory as an emblematic way of accounting for one
of the creative origins of *Sister Carrie;* but beyond this admission, the
passage contains an unhesitating placement of self. His survey of New
York life is conducted not from an elevated place, but from a position
in the midst of destitution and as one of the destitute: "I saw, I admired,
and I resented being myself poor and seeking."[12] He is outside, looking
in and up, as unsheltered and vulnerable as Stieglitz had been secure in
his lofty headquarters. Later he looked at the financial district with a

similar sense of self deprecation and exclusion: "I, having no skill for making money and intensely hungry for the things that money would buy, stared at Wall Street, a kind of cloudy Olympus in which foregathered all the gods of finance, with the eyes of one who hopes to extract something by mere observation."[13] Sight does not provide Dreiser with any equanimity; it originates in desire and it seeks to distance him from his own lowly circumstances almost as if vision were an acquisition or a form of power commensurate with the money and status that he knows himself to lack. "I hoped," he says, "to extract something by mere observation."

Clearly he did not want to remain one of "the first articulate voice[s] from the great poor class of the city," any more than he wanted to write like Mary Antin in *The Promised Land* when she observed that although "the great can speak for themselves, or by the tongues of their admirers, the humble are apt to live inarticulate and die unheard." Where Antin pointed out that she was "born among the simple with a taste for self-revelation,"[14] Dreiser struggled to enter the mainstream culture, to leave the perspective of the poor behind, and to report on powerlessness as if he were not identified by sentiment or experience with that which he described. In these circumstances, seeing empowers by granting the seer an identity distinct from what he observes, but its price is paid in alienation from the circumstances, which originally prompted him to communicate.

It is here that we can understand Steiglitz's most profound influence on the young reporter of 1899; for, with personal motives which remained entirely hidden from Dreiser, Stieglitz nonetheless showed him the picturesque as an approach to the city; he demonstrated that poverty could be an appropriate subject for high art and that one could find a disinterested beauty in the face of it. Dreiser needed both aspects of Stieglitz's work: one removed the stigma of his own poverty and the other mitigated its pain. Had he known the private ends served by Steiglitz's interest in the poor, he might have seen more clearly or at an earlier point the differences in their deepest loyalties; but he did not. He took a formal strategy and, misreading its source, used it to legitimize his own concern with issues of cultural dispossession. For him, seeing became a technique of psychological mobility, expansion, and in ways he could not initially articulate, a strategy of resistance to the meaning of his own originality. If Dreiser could not literally occupy the spacious quarters of the New York Camera Club, he could, at the least, make of his own mind and sensibility a similarly sheltered haunt above the city where the materials of his art could be sifted, composed, and comfortably objectified.

IV

When Dreiser met him in 1899, Alfred Stieglitz was already well established as a champion of the American avant-garde and as a charismatic
figure whose conversation drew many of New York's artists, writers,
and collectors into his presence. The young and the rebellious—those
who felt comfortable in neither the drawing rooms nor the offices of
the time—soon discovered that Stieglitz turned their dissatisfactions
from liabilities into the signs of election to a new kind of spiritual
community.

To Paul Rosenfeld, he was a crusader against crass materialism.
"The machine," he claimed in the Dial, "succeeded during the nineteenth century in reducing most of humanity to spiritual inertia and
caused man to seek to regard objects with only the eyes of commerce
and industry, and not with those of the earth-loving spirit, but Stieglitz
reversed that process and made the machine once more serve the human
spirit."[15]

To others, he came to have a Whitmanesque stature, as if his person
embodied a mythic American potential, the possibility of spiritual
coherence amid widespread industrial waste. Waldo Frank valued his
denunciations of "commercialism ... institutionalism ... hypocrisy,
deceit, conformity, mass values, conspicuous waste and the suppression
of genuine feeling and emotion."[16] The most famous of his little gal-
leries, 291 (established in 1905), acquired a similar mythic identity: "If
ever institution in America came to bring the challenge of the truth of
life to the land of the free, and to show the face of expressivity to a
trading society living by middle-class conventions, it was ... 291."[17]

Stieglitz's appeal seems largely to have grown out of his decision
to bring the problems of convention—in both art and human conduct—
into the foreground of discussion. Neither in life—which he lived as a
bohemian—nor in the art championed by 291 would Stieglitz adhere
to traditional values. When his first marriage to Emmeline Obermeyer
(1893), which he considered to be a dreary, bourgeois arrangement, dis-
solved, he shared a hotel-cafeteria-summer-home existence with Georgia O'Keeffe; his gallery and publications became the voice of new,
predominantly European ideas about the nature and aims of pictorial
representation.

His young friends read his photographs in the light provided by
his opposition to American materialism; they considered that each
image abstracted "the essence of the object as an 'independent thing'
from its obscurantist role as a thing with a use."[18] To them, his work

constituted a serial decoding of the material world, as if, by removing photographs from every context except the blank page or the empty gallery wall, one could strike at the very ideas of utilitarianism, convention, and commerce. They considered his work an insult to a world that rarely offered respect and predictably expropriated resources and people for preordained uses. To some extent, their lionization of him was based on a firm, if intuitive, sense of the political status of Stieglitz's seeming withdrawal from politics: to remove photographic prints from a "vulgar" world—as he did—and to insist on evaluating them as precious art objects—as he insisted—is to envision a less corrupt world to which they more properly belong.

In a sense Stieglitz did with his photographs something analogous to what Hart Crane did in his major narrative poem, "The Bridge." Both artists wanted to find cultural greatness in a land that they loved, but they ended by projecting the promise of America into the distant future, as if America lay underneath its skyscrapers and macadam highways or vaulted over it in the Brooklyn Bridge, potentially vital but encased in values and social priorities which held it immobile, vile, and in need of forgiveness. Crane spoke of city workers as

> O caught like pennies beneath soot and steam,
> Kiss of our agony thou gatherest;
> Condensed, thou takest all—shrill ganglia
> Impassioned with some song we fail to keep.

He posed the bridge, his "myth to God," above and against history:

> Migrations that must needs void memory,
> Inventions that cobblestone the heart,—
> Unspeakable Thou Bridge to Thee, O Love.
> Thy pardon for this history.[19]

Stieglitz saw America in similar terms of vitality as yet unexpressed. He agreed with Marius de Zayas that "American[s] . . . do not see their surroundings at first hand. They do not understand their milieu. . . . The real American life is still unexpressed. America remains to be discovered."[20] His task as a photographer took shape in response to this idealism and disappointment. If Americans were to envision a more noble future, they had first to disengage themselves from the conventions which held them captive, to escape pre-established relationships with the world of historical circumstance, to see anew. Like William Blake cleansing the doors of perception in nineteenth-century England, Stieglitz was a primitive deconstructionist, an artist who attempted to draw the viewer of his photographs out and away from convention as a gesture preliminary to the imagination of something better.

Nowhere was Stieglitz's attempt to avoid historical contingency more clearly articulated than it was in the work of his final years, an extended meditation on clouds, called "Songs of the Sky" or "Equivalents." To H. J. Seligman in 1934, Stieglitz spoke of them as the culmination of his career as a photographer, for they embodied all that was implied more indirectly in his earlier work. They were, he said, extremely "democratic." "Through clouds to put down my philosophy of life—to show that my photographs were not due to subject matter—not to special trees, or faces, or interiors, to special privileges—clouds were there for everyone—no tax as yet on them—free."[21] Since clouds were available to everybody, the beauty of photographing them could not come from special access to an artistic subject, but must reside in the method of seeing in and of itself. "I have found," he said, "that the use of clouds in my photographs has made people less aware of clouds as clouds in the pictures than when I have portrayed trees or houses or wood or any other objects. . . . The true meaning of the 'Equivalents' comes through without any extraneous pictorial factors intervening between those who look at the pictures and the pictures themselves."[22]

Stieglitz's effacement of the subject was, then, a movement made in the face of criticism and in response to his growing sense that the genius of his photography lay not in his subject—which could be considered to be privileged (as, for example, when he photographed Georgia O'Keeffe in intimate circumstances), but in his ability to render response and counter-response in terms of light and dark and abstract form. "Shapes in relationships" were what, at base, held his attention in this sequence. They expressed something entirely private and, he hoped, entirely sharable: "What is of greatest importance," Stieglitz said, "is to hold a moment, to record something so completely that those who see it will relive an equivalent of what has been done."[23] It was in this way that Stieglitz hoped to effect change, if only by drawing his viewers out of traditional modes of seeing and evaluating into another, more direct, less ossified or preconceived relationship with the visible world. He used the camera as if it could, merely by being "straight," by presenting so unmediated a record, cleanse perception in the most extended and profound way.

The roots of this kind of thinking about the relationship of form and subjectivity must certainly be found, at least in part, in the discussions of modern art that appeared in the magazine that Stieglitz published, *Camera Work*. Stieglitz worked very much in context and in the midst of debate about both the genesis and affects of contemporary art.

Two essays come immediately to mind: Charles Caffin's article on Matisse (1909) and Marius de Zayas's piece on the early work of Picasso (1911). These critics, both friends of Stieglitz, struggled to understand and to explain the modernist impulse to retreat from the replication of the empirical world. Caffin observed that Matisse worked "not from optical data but from a 'mental impression,'" that he did not value observation for its own sake, but that he tried to "interpret the feeling which the sight of the object stirs in him."[24] Two years later Marius de Zayas interviewed Picasso, who was then in his Analytic Cubist period, and concluded that the artist "receives a direct impression from external nature, he analyzes, develops, and translates it, and afterwards executes it in his own particular style, with the intention that the picture should be the pictorial equivalent of the emotion produced by nature."[25]

"The pictorial equivalent of the emotion produced by nature"— Picasso's description of his own work—is almost exactly the language Stieglitz later used to explain the meaning of his "Songs of the Sky." He considered that the given world, captured by the lens in a certain way, expressed the interior, emotional world of the photographer in the same way that an expressionist painting revealed the artist: "My photographs are a picture of the chaos of the world and of my relationship to that chaos. My prints show the world's constant upsetting of man's equilibrium and his external battles to reestablish it."[26]

Like his peer, Jacob Riis, Stieglitz was never a systematic thinker, but all of his comments about things like "touch" or "chaos" in a photograph proceed from a certain posture before the physical world, a posture in which the boundaries between the ego and its object of reflection are blurred so that a momentary union between seer and seen seems to be achieved. This is certainly what Herbert Seligmann concluded when he visited him in 1925: "He drew a diagram making adjacent dark spots. 'Here,' he said, 'is reality. When that is seen it is so close to the mystical that the dividing line is almost imperceptible.' He drew a line between the two spots. 'That is the line of my life running between them.'"[27]

We can find this mystical thread running through many of Stieglitz's self-reflections. He considered himself to be emptying out into a world waiting for his acknowledgment, and to solve through this receptivity two correlated problems: the burden of his own isolated ego: "When I am no longer thinking, but simply am, then I may be said to be truly affirming life"; and the liberation of the subject: "[If you permit yourself] to be free to recognize the living moment when it occurs, and to let it flower without preconceived ideas about what it should be then you are the moment."[28] Of these two aesthetic goals, it was the

second that his friends and colleagues considerd to be the most important and it was from this "liberation" that they derived the social meaning of Stieglitz's work, seeing in his lack of preconceived ideas both an
act of homage to the subject and an affront to capitalism's tendency to
value things according to their usefulness as items of exchange.

The other side of this withdrawal of the image from its context is
that negations or stripping away can never be politically or socially
neutral. In Stieglitz's own eyes, his images were not withdrawn absolutely but only withdrawn into a different and more private system of
signification. Whatever his colleagues might have believed about his
photographs' implicit rebuff to commerce, Stieglitz was concerned with
more than absence or assault. The images expressed something deep and
positive as well; they were signs of his internal identification with certain configurations given in the empirical world and they recorded feelings that he wanted to share with or evoke for others. In the "Equivalents" sequence, he used his lens to identify clouds with chaos. In an
earlier series of photographs of Lake George, he spoke of the meeting
of mountains and sky in terms of gentleness and touch. In still earlier
work, he identified the poverty of American immigrants with vitality
and spiritual health. Always, Stieglitz said, "I decided to photograph
what was within me."[29]

But he must have suspected the precariousness of this strategy as an
act of communication, for the price of this type of ideology, this withdrawal of meaning into the private and abstract, was the loss of the
photographs' importance as reports or as vehicles of intersubjectively
sharable meaning. The meaning of America, as presented by Stieglitz's
photographs of America, no longer resided in the appearance of America, but in some mythic, invisible place—either inside the psyche of the
artist or in the abstracted ideas of his friends—so that the withdrawal
of images from America's commercial landscape was really a way of
drawing the images into some other scheme of primarily autobiographical relevance. J. C. Rowe would say that such artistry "serves the dominant ideology . . . in terms of a compensatory substitution of its own
form for failures or omissions of the culture."[30] In other words, Stieglitz's assault on one kind of cultural authority disguised the authority
of his own art, his own control of the meaning of images so that work
that was begun as protest against the hegemony of a commercial culture
turned, finally, into a repetition of the structures of cultural meaning
that seemed initially to be under attack.

To attribute to Stieglitz's schemes a "lack of context" and to speak
of his avoidance of the historical contingencies of his time would be to
misjudge them, for they were acts of expropriation continuous with,

analogous to—not the opposite of—the expropriation of men and resources of the business community that Stieglitz so despised. It would also miss the cultural situation in fin-de-siècle New York City, where the individual artist pitted himself against bank vaults and felt his own ego to be the only other viable repository of cultural meaning, as if the imagination were all there was to assert against granite, stone, and the distress of human inconsequence in the city. If America's appearance did not have a public, sharable meaning aside from commercial utility, it had only the obscurity, the illegibility of the private psyche.

And illegible it was. When Theodore Dreiser looked at Stieglitz's work, he saw none of Stieglitz's equation of forms with internal states of emotion; he saw forms, he saw what the photographs reported. For him the excitement of discovering Stieglitz lay in seeing that street life could become an admissible subject of high art. Stieglitz's choices seemed to legitimize what Dreiser himself saw of American life—the underside of success, the price of struggle, the shabby consequences of greed and aggression and neglect on the streets of America's otherwise fine and glowing cities.

It was Stieglitz's early work—pictures of New York City in the late 1890s—that Dreiser had seen when he interviewed Stieglitz in the New York Camera Club. Stieglitz may have been living on Madison Avenue and frequenting a prestigious artists' club, but he was also wandering the same streets that were part of Jacob Riis's newspaper beat, responding in his own way to Mulberry Bend, Hester Street, and all of the vilified haunts that appear in Riis's famous book, *How the Other Half Lives*. But Stieglitz saw and experienced an "other half" that would have mystified Riis. If we take quite seriously Stieglitz's assertion that "Equivalents" made explicit a subjectivist movement that was implicit in his work from the beginning, we can discover a reason for this divergence of pictorial treatment. Where Riis found others in distress, recognizing their destitution as an analogue of his own earlier disinheritance as an immigrant, Stieglitz found the poor vital and honest in proportion to the stuffiness that characterized the bourgeois life to which he was born:

> Nothing charms me so much as walking among the lower classes, studying them carefully and making mental notes. They are interesting from every point of view. I dislike the superficial and artificial, and I find less of it among the lower classes. That is the reason they are more sympathetic to me as subjects.[31]

One kind of alienation played into the other, as if the disinheritance of the aesthete could find expression only in pictures of the literal disinheritance of the immigrant poor. Stieglitz had photographed the streets, the buildings of the city, its activity, and its weather in the same way that he later photographed clouds. Always, he said, "there seemed to be something closely related to my deepest feelings in what I saw, and I decided to photograph what was within me."[32]

A similar attraction existed when he photographed "The Steerage," a photograph (his favorite) of the stratified life of an ocean passage to Europe in 1907. To those familiar with the history of photography, the circumstances surrounding its creation are well-known, but the story is important enough to repeat:

> [My] wife insisted upon going on the *Kaiser Wilhelm II*—the fashionable ship of the North German Lloyd at the time. Our first destination was Paris. How I hated the atmosphere of the first class on that ship. One couldn't escape the *nouveaux riches*.
>
> I sat much in my steamer chair the first days out—sat with closed eyes. In this way I avoided seeing faces that would give me the cold shivers, yet those voices and that English—ye gods!
>
> On the third day out I finally couldn't stand it any longer. I had to get away from that company. I went as far forward on deck as I could. The sea wasn't particularly rough. The sky was clear. The ship was driving into the wind—a rather brisk wind.
>
> As I came to the end of the deck I stood alone, looking down. There were men and women and children on the lower deck of the steerage. There was a narrow stairway leading up to the upper deck of the steerage, a small deck right at the bow of the steamer.
>
> To the left was an inclining funnel and from the upper steerage deck there was fastened a gangway bridge which was glistening in its freshly painted state. It was rather long, white, and during the trip remained untouched by anyone.
>
> On the upper deck, looking over the railing, there was a young man with a straw hat. The shape of the hat was round. He was watching the men and women and children on the lower steerage deck. Only men were on the upper deck. The whole scene fascinated me. I longed to escape from my surrounding and join these people.
>
> A round straw hat, the funnel leaning left, the stairway leaning right, the white draw-bridge with its railings made of circular chains—white suspenders crossing on the back of a man in the steerage below, round shapes of iron machinery, a mast cutting into the sky, making a triangular shape. I stood spellbound for a while, looking and looking. Could I photograph what I felt, looking and looking and still looking? I saw shapes related to each other. I saw a picture of shapes and underlying that the feeling I had about life. And as I was deciding, should I try to put down this seemingly new vision that held me—people, the common people, the feeling of ship and

ocean and sky and the feeling of release that I was away from the mob called the rich—Rembrandt came into mind and I wondered would he have felt as I was feeling.

Spontaneously I raced to the main stairway of the steamer, chased down to my cabin, got my Graflex, raced back again all out of breath, wondering whether the man with the straw hat had moved or not. If he had, the picture I had seen would no longer be. The relationship of shapes as I wanted them would have been disturbed and the picture lost.

But there was the man with the straw hat. He hadn't moved. The man with the crossed white suspenders showing his back, he too, talking to a man, hadn't moved, and the woman with the child on her lap, sitting on the floor, hadn't moved. Seemingly no one had changed position.

I had but one plate holder with one unexposed plate. Would I get what I saw, what I felt? Finally I released the shutter. My heart thumping. I had never heard my heart thump before. Had I gotten my picture? I knew if I had, another milestone in photography would have been reached, related to the milestone of my "Car Horses" made in 1892, and my "Hand of Man" made in 1902, which had opened up a new era of photography, of seeing. In a sense it would go beyond them, for here would be a picture based on related shapes and on the deepest human feeling, a step in my own evolution, a spontaneous discovery.

I took my camera to my stateroom and as I returned to my steamer chair my wife said, "I had sent a steward to look for you. I wondered where you were. I was nervous when he came back and said he couldn't find you." It told her where I had been.

She said, "You speak as if you were far away in a distant world," and I said I was. "How you seem to hate these people in first class." No, I didn't hate them, but I merely felt completely out of place.[33]

When Stieglitz printed the image, he divided the picture into imaginative halves: in his photograph, upper and lower classes were represented spatially—and though Stieglitz himself disavowed the importance of the photograph's spatial division, to others its bifurcation was its most salient feature. "The first person to whom I showed 'The Steerage,'" Stieglitz remembered, "was my friend and coworker, Joseph T. Keiley. 'But you have two pictures there, Stieglitz, an upper one and a lower one,' he said. I said nothing. I realized he didn't see the picture I had made."[34]

In this dialogue—which records Stieglitz's dismay—may be read the central drama of his relation to New York's poverty—at least as it is represented metonymically by this shipboard world. For Keiley saw what was reported in the picture: the place of the poor in the lower hold, and the position of the other passengers in the open air where they could choose to look either to the sky or to the activities below and the gangplank that separated one from the other. Stieglitz did not see darkness or enclosure but only the coincidence of his own private, half-hidden motivations and a tableau that could express them. He did not escape into the open air but into the vitality he had imaginatively bestowed upon these less fortunate travelers. As the clouds expressed "chaos" so the poor expressed freedom and authenticity. As he tended frequently to do, Stieglitz took the reported meaning of the physical world and collapsed it into autobiography; he made the visible into a cipher for his own psychological proclivities. This expressionism ended by achieving the opposite of what Stieglitz thought it achieved: it expressed not homage, but disregard, and it sustained a relation of power, a structure of domination within the camera-subject nexus, for such a posture takes the world apart from itself and once again subordinates it to someone else's intentions for it. In this one example, the photograph's meaning is independent of the condition of the subjects; it does not report *their* freedom, but Stieglitz's; it does not express them, but him—and it does this even though Stieglitz quite earnestly felt a mystic union with the vision through his viewfinder. What Elizabeth Bowen has written in *Death of the Heart* is relevant to Stieglitz as well: "Like an empty room with no blinds, his imagination gapes on the scene, and reflects what was never there."[35]

This cacophony of circumstances makes "The Steerage" an important and disturbing as well as a beautiful photograph, and it suggests the extraordinary complexity of the issues raised by Stieglitz's photography in general. For this image may tell us very little about those in the steerage class and it may evoke no equivalent emotion in an audience, but it does indicate that Stieglitz was drawn to them and not to something else. In the midst of all his disclaimers about the irrelevance of subject matter and the supremacy of beautiful form in pictorial photography, subject matter was nonetheless centrally important to him. Why should he, who was so obviously disinterested in poverty in the way that Riis was, choose poverty as his subject? To the extent that Stieglitz tried to elucidate his own motives, he simply made poverty the repository of all that bourgeois life was not: he tried to believe that by seeing vitality, he could join it; by criticizing the bourgeoisie, he could leave it; by refraining from a commercial career, he could find himself in a predicament equivalent to that of the working people he imagined.

Left at this level, Stieglitz's obfuscation of issues is apparent. Hawthorne and James knew quite clearly that life could not be appropriated through vision; it had to be lived. But if we meditate on what Stieglitz was living, as opposed to the life of freedom he desired to live, we can understand these pictures as equivalents of sorts: they tell us a great deal about Stieglitz's struggle to understand the social restratification implied by his commitment to photography and by his abandonment of Madison Avenue and Wall Street. He was not entirely mistaken about the social structure which gave to artists and photographers a marginality similar to that experienced by Riis's alien poor. If I were to name the "equivalent" expressed by Stieglitz's "poor" photographs, it would have nothing to do with "freedom"—which I regard as a false vernacularism or a misreading of the characteristics of class—but with this common experience of peripheral status in a world obsessed with money. As Edith Wharton put it in *The Age of Innocence,* her novel about New York at the turn of the century, "I don't know that the arts have a milieu here, any of them; they're more like a very thinly settled outskirt."[36]

In saying this, I am not trying to equate Stieglitz's self-selected bohemianism and the actual destitution of New York's poor, but the convergence of bohemianism and the city's ugliness as a subject of attention is crucial: it explains why Stieglitz, as well as Riis, is of pivotal importance in understanding the emerging representational strategies of the literary realists of his generation. In the pursuit of entirely private, expressionistic goals, Stieglitz turned his camera on the downtrodden; in the interests of establishing photography as a fine art, he hung such pictures on gallery walls alongside photographs such as those of Gertrude Kasebier and F. Holland Day who cherished, among other things, the home and the sanctity of childhood and womanly purity. Had he been destitute himself, he probably could not have done this with so little struggle; but as an errant son, as the child of an established manufacturing family who was "innocently" if eccentrically engaged in self-expression, he has was in a position to elevate poverty into a legitimate subject for high art.

Critics like Paul Rosenfeld adored him for this reason:

> Stieglitz has shoved the nozzle of his camera into hells where man's hand was rested cruellest, and caught filthy smoke and grimy skies, iron and cinders and strong steely wires.... He has portrayed the people of the American streets and of the workmen's cottages, of the kitchens, the slums, the studios, the little westside apartments, and the shining limousines.... And car-horses steaming amid smudged New York snow have made pulse the light no less lambently than whitest cloud shapes breezing through shoreless blue; cloth of an old coat no less burningly than the singing downtown skyline; brick

walls torn and brick walls laboriously upbuilding no less rapturously
than slender fingers licked with sensitive life.[37]

Waldo Frank also adored him:

> The true analogue to Stieglitz with his camera before a patch of sky,
> a city street, a face, a tree, a hand, is . . . Blake recording his poem to
> the Tyger . . . all that occurs to the man and with the man is focused
> (photographed one might say) by the lens of his acceptance of life.[38]

Without apparent concern for the social implications of his photographs, Stieglitz was able, in fact, to find railroads, skyscrapers, bones,
clouds, and poverty equally picturesque. He nonetheless affected the
nation's opinion of what was an admissible repertoire of subjects for the
fine arts. Riis had turned his camera on poverty in the name of science;
Stieglitz had photographed poverty in the service of self-realization and
in the context of European expressionism. Dreiser took from Stieglitz
his dedication to the picturesque, his interest in powerlessness as a subject for high art, and added it to the perspective of the "other half"
whose deprivations and desires Stieglitz could never, with his dedication to self-expression, have surmised.

V

On 8 December 1915, the Nation carried an attack on Dreiser by Stuart
P. Sherman of Harvard University. The occasion was provided by the
publication of Dreiser's new novel, The Genius, but Sherman used this
event to criticize a more widespread and, to him, dangerous philosophy
of literary representation.

> A realistic novel no more than any other kind of novel can escape
> being a composition involving preconception, imagination and divi-
> nation. Yet, hearing one of our new realists expound his doctrine,
> you might suppose that writing a novel was a process analogous to
> photographing wild animals in their habitat by trap and flashlight.
> He, if you will believe him, does not invite his subjects, nor group
> them, nor compose their features, nor furnish their setting. He but
> exposes the sensitized plate of his mind. The pomp of life goes by,
> and springs the trap. The picture, of course, . . . simply re-presents.
>
>   The only serious objection to this figurative explanation of the
> artistic process is the utter dissimilarity between the blank impartial
> photographic plate . . . and the crowded, inveterately selective human
> mind.[39]

The issues that were joined by Sherman's comment about literature and photography were more complicated than he knew, and it is ironic that in 1915 Dreiser would probably have agreed with his view that writing should not simply parrot back something given in the empirical world. If Stieglitz had taught Dreiser anything, it was certainly that the viewfinder of the camera—or the mind used as a viewfinder—was the vehicle of active composition. It was just that he wanted, like Stieglitz, to compose from the raw and desperate side of urban, industrial life as well as from its more genteel characteristics. Since he considered American culture as something which generally denied the expression of large areas of experience (Dreiser recalled thinking that "life as I saw it, the darker phases, was never to be written about. Maybe such things were not the true province of fiction anyhow"),[40] he did write in rebuke to Sherman's sense of decorum, but he did not want simply to re-present the conditions of life as if he had no sensibility at all. Even in his earliest newspaper sketches he turned the city into a Stieglitz-like spectacle for himself, making the ugly the vehicle of beauty and himself as seer, as spectator of the beauty, into someone capable of a separate and effective destiny. Paul Rosenfeld's appreciation of Stieglitz's accomplishment—"There is no matter in all the world so homely, trite, and humble that through it this man of the black box and chemical bath cannot express himself entire. A tree, a barn, a bone, a cloud, have released the spirit in Stieglitz"[41]—recalls Dreiser's own explanation of Stieglitz's methods—"He will look for pictorial possibilities in the landscape. The bare, hard outlines of a scene are nothing in themselves. It is the beauty which he sees."[42] Indeed, these summaries of the achievement of a certain mode of photography can serve equally to describe Dreiser's rendering of similar subjects and scenes. In another article written for *Success* in 1910, he wrote:

> I sometimes look at this factory in the glow of the evening sun or through the pleasing drizzle of a rainy day, and ponder over the philosophy of it. Its stature is so graceful; its tall chimney so artistic. If you view it as a picture solely, it is one of those satisfying anomalies which the heart of man greatly rejoices in. The color of its walls is so adequate, the height and form of its chimney so wholly pleasing and ornate. Seen through the glow of the setting sun or when the rain or fog are softening it to a mere shadowy outline, it appears a sweet concoction of Nature, like a flower or a tree. The beauty of it makes it agreeable to that idea.[43]

Of the New York car yard he said:

> If I were a painter one of the first things I would paint would be one or another of the great railroad yards that abound in every city. . . .

Only I fear that my brush would never rest with one portrait. There would be pictures of it in sunlight and cloud, in rain and snow, in light and dark.[44]

Even when he approached the homes of the factory workers and meditated more generally "On Being Poor," Dreiser saw his subjects in terms of the disinterested spectacle they presented.

And yet, shabby and depressing as are these facts, there is a collective, coherent charm and color about the effort itself which to one who views it entirely disinterestedly is not to be scoffed at.... [Life] can and does achieve an aesthetic whole—beauty no less.... You and I may argue that rats and flies and bedbugs are not aesthetic and join no aesthetic whole, but examine more closely with lens and concentrated interest of the mind and its response to organization and effort and then judge.[45]

He had good reason to hold firmly to this kind of disinterest, for at times it seemed to him that aesthetic judgment was all that stood between him and utter destitution. In 1902, two years after the failure of *Sister Carrie,* his first extended foray into fiction's wider province, Dreiser collapsed. Exhausted, without resources, literally subsisting on food gathered from the streets, he joined Hurstwood and the other bums of his own imagination in a struggle to stay alive.

His brother, Paul Dresser, the noted songwriter and music publisher, helped him by footing the bill for his stay at the Muldoon Hygienic Institute in Purchase, New York. Curiously, Dreiser's own contribution to restoring his health was to do manual labor for the New York Central Railroad in a small shop in Spuyten Duyvil, New York. It was as if he had taken Stieglitz's romanticism about common men and honest labor to heart, agreeing that such people possessed a vitality of spirit denied to those of a more intellectual bent.

If the impulse to join the working class originated with a Stieglitz-like hope, Dreiser's own experience on the job quickly demystified the enterprise. In a series of autobiographical fragments written between 1902 and 1904, Dreiser admitted the vanity of his original perspective: "It had always been my fancy that the workingman ... was after all the happy man.... Here I found that there was no such ideal in existence."[46] The work was hard and monotonous; fear of being laid off lingered over everyone; boredom was constantly interlaced with anxiety. When he was honest with himself, he knew unquestionably that he had entered "the dreariest [period] of [his] life. It combined the various qualities of sickness, want, friendlessness and limitation."[47] As if to dissociate himself from this misery, he began to act like the observer of his own experience: "I seemed to ... become two persons. One of these was a tall, thin, greedy individual who had struggled and thought always for himself and how he could prosper, but was now in a corner and could not get out, and the other was a silent, philosophical soul who was standing by him watching him in his efforts and taking an indifferent interest in his failures."[48] Repeatedly he drew on what he knew of Stieglitz's skill in framing and finding interest in the coalescence of rough-hewn details. "It was depressing to me, I'll admit, and yet it had the element of charm, which comes to one who is interested in the spectacle which his fellow man presents."[49] The poor, in his opinion, were those who lacked the ability to discern the picturesque. Though he later came to think of the problems of the working class as a function of information purposefully withheld from them by an established power structure, at this early point in his life, it was not class distinction but a kind of déclassement which appealed to him about the status of the artist. "Without money, and at times with so little that an ordinary day laborer would have scoffed at my supply, I still found myself meditating gloomily and with much show of reason upon the poverty of others ... but poverty of mind, of the understanding, of taste, of imagination—therein lies the true misery, the freezing degradation of life."[50]

Yet even as he used the aesthetic sensibility that he thought appropriate to the artist, Dreiser was gathering evidence in Spuyten Duyvil that would eventually lead him to challenge the adequacy of Stieglitz's way of seeing. For one thing, the workshop was not set out as a view or spectacle to be taken in; Stieglitz might look at the scene as if it were a beautiful panorama, but for Dreiser it was a place of interaction. The more he drew back to watch himself and his fellows, the more inscrutable they became; they could, he said, "detect at once the mental interloper."[51] Like John Berger's French peasants, Dreiser's workers lacked the perspective that might lead them to evaluate the quality of their experiences; they could not act like "a crow on the tree watching." To most of them, Dreiser said, "it was all but meaningless. Having labored on but portions of it, they could scarcely conceive it as a whole."[52]

By the same token, Dreiser's experiences at Spuyten Duyvil also showed him the fallacy of thinking that a supervisory view presented things more accurately. To the extent that Dreiser became a real worker immersed in the scene, he considered that he became invisible to others in positions of authority: "What did they know of this underworld which they ruled; what did they know of themselves. Blind leaders of the blind, strutting about the world.... What did they care for me.... I was nothing to them. They did not see me."[53]

In certain ways, Dreiser's experiences at the New York Central workshop were the decisive experiences of his career as a narrative writer. He had, in essence, entered the frame of his own imagination, an act which dispelled the mythic relationship between beauty and poverty and exploded any ideas Dreiser may have harbored about the self-understanding of the working class. He concluded that no one in the body politic saw any one else very clearly, and I think this dilemma was intimately connected to his movement away from picturesque treatments of subjects to a much more complex handling of the problems of exposure, secrecy, and concealment in narration. He came to think about realism as a mechanism of disclosure, as something asserted against the culture's tendency to conceal its true strategies of power.

Having seen the picturesqueness of phenomena, having used such detachment to assure himself an almost mystical immunity from poverty's possible damage to himself, he nonetheless knew life to be more complicated than such an attitude of disinterested surface concern could render. If Stieglitz's example had helped to authenticate the look of poverty as an appropriate subject for art, it did not, finally, provide a full model for Drieser's relation to the most problematic subject of his career: "Poverty is not desirable," he said, "Its dramatic aspect may be worth something to those who are not poor, for prosperous human nature takes considerable satisfaction in proclaiming: 'Lord, I am not as other men,' and having it proved to itself. But this thing, from any point of view is a pathetic and disagreeable thing."[54] Dreiser came eventually to see that Stieglitz's method of photographing reinscribed a system of subordination and that his photographs were acts of cultural definition that occurred above-the-heads-of or outside-the-lives-of the people who provided the image. To the mature Dreiser, Stieglitz would remain a bohemian dabbler in a class with Richard Harding Davis, Hobart Chatfield-Taylor, and Percival Pollard—"fellows who move in society when they choose and out of it when they don't," men who dropped in and out of Mulberry Bend and the Bowery as "flavor."[55]

Dreiser pirated his own labor diaries for certain episodes in The "Genius," published in 1915. Several years later, in 1923, he seems to have returned to the same material with a heightened sense of what was contained in it. Joseph Coates, friend and editor, responded to Dreiser's suggestion of a new book of nonfiction in a letter on 28 September: "What you say about a book of personal experiences interests me, and naturally I think of your work on the New York Central Railroad among those who toil with the hands. . . . If your personal experiences are in similar fields to those that Jacob Riis has browsed in, I fancy they might do quite well in the shape you have first cast them as to put them into fiction."[56] Coates's comparison of Dreiser's proposal to the work of Jacob Riis was a comparison that almost any literate New Yorker might have made at the time. Even though *How the Other Half Lives* (1890) had been published over thirty years before, it remained the classic account of the haunts of poverty as seen through the eyes of the American middle class. Coates joined the two artists in imagination because they both had "browsed in" the life experiences of people who were not usually represented in cultural discourse.

But for Dreiser there was a congruence deeper than subject matter in this comparison, for he came to understand that Riis's approach to the Lower East Side was in some sense an answer to the questions which increasingly pressed upon him: What lay underneath? What was hidden? What structures accounted for the details in flux? For Dreiser, art increasingly became a search for withheld secrets; appearances became the foil of a more complex and sinister structural reality. He could look at loveliness and remain uneasy in a way that Stieglitz never did. In a sketch called "The Beauty of Life," he said that "every aspect of the scene reveals something pleasing which could scarcely have been the result of a cruel tendency, and yet we know cruelty exists, or ... at least a tendency to contention ... and this is not what is generally revealed in any scene."[57]

He grew to see the novelist's duty as a commitment to revealing "what is [not] generally revealed in any scene" and to connect this duty with a more comprehensive theory of cultural control and, ultimately, of cosmic organization. In his eyes, "the social mystery, the lurking human secret" did not have to do with how the poor lived, as it did for Henry James, but rather with a search for the parameters of social organization that erased the traces of its own authority even as it maintained and disguised the suffering and inequality it promoted. His understanding of the American political scene and his later groping theories of the universe and the forces which control the fates of men are interesting because they reinforce each other structurally; they are parallel systems whose explanatory schemes seem to have appealed to him because they could account for the powerlessness and invisibility that Dreiser saw accorded to his own life and to those like himself. What John Berger has said of "clumsy" primitive art is equally true of some of Dreiser's writing: "For what it is saying was never meant, according to the cultural class system, to be said."[58]

One of Dreiser's major contributions to literature can be considered his ultimate resistance to Stieglitz's formulaic rendering of the world as something arranged for the view of a privileged onlooker. His decision to yield, to write about the world as it looked from the unprivileged point of view, is what called forth Stuart Sherman's vehement attack, and it is what eventually strengthened his own animosity toward the world of high culture that Sherman represented to him. Dreiser's major complaint about "Life, Art and America" remained remarkably consistent throughout his career, and it was, put in simple terms, a complaint against the cultural misconstructions that ensued from a limited field of vision, against the distortion that was the inevitable concomitant of the overly restricted frame, and finally it took the form of self-criticism for having once assumed that perspective himself.

Riis offered Dreiser an alternative model of social understanding because he used the camera in unorthodox places and according to certain compelling personal priorities which joined private to public concerns, combining artistic practice with a supposedly scientific theory of vision. It was not simply that Riis was honest but that he was honest about what lay behind the tenement walls that hid so much turmoil. Riis went inside with the camera, he gained access, he revealed what the culture at large would have been more comfortable either ignoring or stereotyping or dismissing in other more damaging ways.

VI

In fact, it was to assay the attitudes and misperceptions of Stuart Sherman's "type" that Jacob Riis, a rather evangelical newspaper reporter, himself a Danish immigrant, had taken the camera into the slums and inside the buildings that Henry James, on his American tour, had mistakenly considered to be so highly evolved. In his autobiography, *The Making of an American* (1904), Riis remembered, "[T]he whole idea of that reform was to better the lot of those whom the prosperous up-town knew vaguely only as 'the poor.'" His efforts had begun a full twenty years before James roamed the American continent, but his audience was James's audience, and he thought his perspective to be more informed.

Where those like James visited New York and stayed in elegant hotels or in the private homes of friends, Riis had come to America as an immigrant and had suffered from the conditions he spent his later life trying to rectify. Curiously it was he—a man whose empathy had roots in personal experience—who helped to ossify New York's social multiplicity into the categories "we" and "they" by publishing a highly influential article for *Scribner's Magazine* (1889) called "How the Other Half Lives."

Riis's magazine article grew, by request, into a full-length volume (1900) illustrated by photographs of poorly clothed people, over-crowded streets, and decrepit houses. The book served as a turning point in Riis's life and as a significant event in the history of photography, for it drew the camera into a central position in the social reform move-ment of the time, making it part of the debate about the most proper aims and procedures of the nascent social science movement.

His reliance on the camera arose initially from a simple motive: he was afraid of being ignored. "I wrote," he said, "but it seemed to make no impression." As Charles Loring Brace, who was one of Riis's impor-tant precursors, noted, it was essential to keep the problems of the urban poor consistently under public scrutiny. "I made it a point," Brace recalled, "from the beginning to keep our movements, and the evils we sought to cure, continually before the public in the columns of the daily journals. Articles describing the habits and trials of the poor, edi-torials urging the community to work in these directions, essays dis-cussing the science of charity and reform were poured forth incessantly for years through the daily and weekly press of New York."[59] But by the same token, a barrage of writing about squalor and the need for moral action could stultify itself; people could easily become satiated with the issues of reform. Riis's use of the camera arose from an effort to preclude public lethargy and to establish his discourse in the world of "facts." Not only did the photograph substantiate his case against the tenements, but it also, as Riis put it, "helped to demolish the barriers erected by society against its nether life."[60] In a world not yet inundated with photographic images, pictures of slums could be more effective than charts, statistics, and moral exhortation. In a world that contained both "ignorant poverty" and "ignorant wealth," Riis chose ignorant wealth as his audience.

He was not alone. Although Riis was unschooled in matters of social investigation, and although he never became a systematic thinker, his book was nonetheless published within a framework of competing thoughts about the aims and procedures of social theory and social practice. In fact the relationship between knowledge and its prac-tical application was an issue of foremost importance in the newly established American Social Science Association (ASSA). In its first years (1865–85), the organization had been little more than a coalition of reformers who recognized that modern industrial life had created problems demanding more systematic attention than any single relief program could provide. Coming from disparate backgrounds, the orig-inal members had no common pattern of training, no code of ethics, and no unified social mission. All of them perceived the need for more organized knowledge, and they clung, consequently, to a Baconian faith that amassing information, piling up data, would lead to the discovery of truth. But it was precisely this faith in accumulation and classification that led other empiricists to criticize the ASSA: in their eyes, the ASSA allied knowledge too closely with advocacy and information too closely with reform. Herbert Spencer was one of those most critical of social science as it was used in the service of legislation and particular welfare programs. His most outspoken American colleague was Edward L. Youmans, the editor of *Popular Science Monthly,* who in turn accused the amateur social scientists of "deliberately attempting to deceive the public by cloaking their reformist schemes with the dignity of science."[61] One of his editorials from 1874 exemplifies his position:

> [W]e are of the opinion that it falls short of what should be its chief duty. So far from promoting social science, we should rather say that social science is just the subject which it particularly avoids. It might rather be considered as a general reform convention. It is an organization for public action, and most of its members, not with the impulses of philanthropy, are full of projects of social relief, amelioration, and improvement. Of pure investigation, of the strict and passionless study of society from a scientific point of view, we hear but very little. The Association seems to be but little in advance of an ordinary political convention.[62]

Youman's criticism is important because of the nature of its oppositions: against most of the efforts of his ASSA colleagues, he pitted what he called "pure investigation" and "the strict and passionless study of society." His ideal was to attain in social study the objectivity that characterized the natural sciences. He wanted to take the study of social behavior out of the moral arena and to discredit explanations motivated by views of how people ought to live together.

The point here is not to enter into a debate about the philosophical implications of various social explanatory schemes (or to argue that policy science is never neutral[63]), but to express the poles of opinion about means and ends that characterized studies of the unassimilated people of New York City in Jacob Riis's lifetime. For the political model that underlies each investigative strategy—whether it be the positivistic or the moral exhortatory model—is a dominant/submissive one; that is, each methodology assumes the passivity of the subjects of inquiry in the face of a more powerful group of social actors who will make policies on their behalf. One procedure imagines the actors to be experts who have gathered and evaluated information in an unbiased way; the other envisions the activity of concerned or "enlightened" citizens; but in each case, one of the consequences of the very structure of inquiry is that actual power relations are kept inaccessible to analysis by those whose lives are most directly affected by policy decisions. And, indeed, this is an ignorance that Riis helped foster by writing and photographing the New York slums in the manner that he did.

To make this claim is not to identify Riis's own self-evaluation; but it is correct to say that he understood his book to be part of an increasingly larger body of literature concerning the conditions of the poor or "nether" classes. In his autobiography, he placed himself in the company of Charles Booth (1840–1916) whose *Life and Labour of the People of London* (1891–1903, 17 vol.) was a monumental study of class relations in Victorian England, and Charles Loring Brace (1826–90), another American reform writer whose *The Dangerous Classes of New York* (1872) stands in implicit dialog with Riis's text.

In some ways, Riis's "message" in *How the Other Half Lives* was not substantially different from Brace's earlier conclusions: both men turned their attention to the homeless and the poorly housed; both indicated the greed and neglect of the landlord class; both attempted to explain the causes of a highly visible financial disequilibrium in the city—on a large scale, in the lives of all foreign-born people, and also on the level of individual life, where wage labor, even of the most dedicated and intense, could barely purchase the things necessary to sustain a family. Brace talked about the "thousands on thousands in New York who have no assignable home and 'flit' from attic to attic and cellar to cellar."[64] Riis vilified the tenements, the police flophouses, the very streets and boxes that harbored vagrants. But beyond a common subject and a common good intention, Brace and Riis did not, in the end, remain similarly motivated. Brace saw every piece of philanthropy in the context of a comprehensive social program. Each suggestion in his book was, in effect, a suggestion for changing or reconstituting the character of those in the classes whose belligerence he most feared. Brace imagined the poor as a vast hoard of unrestrained paupers whose raucous behavior might violate a system of unquestioned and unquestionable legitimacy. He approached them as one might approach a pest, as a problem of control. Christianity was his major tool, for he considered that it encouraged voluntary concurrence with established values. In fact, the controlling metaphor of the book and of Brace's imagination is confinement: since he locates the source of social distress so firmly within the character structure of the poor, he always puts them *into* the right frame of mind and then puts them bodily *into* dorms, reading rooms, Sunday schools, factories, and servants' quarters. Brace's solutions call to mind Michel Foucault's distinction in *Discipline and Punish* between punishment as a spectacle revealed, enacted by the representatives of authority on the body of the guilty, and punishment as it is in a modern disciplinary state, as something enacted behind walls, the guilty hidden from the public but visible to and thus constantly vulnerable to an unseen, central authority. Brace worked to divide aggregates of suspicious people into manageable subgroups and then to place them behind closed doors to subject them to the authority of church, state, and private capital. His book, so totally devoid of statistical or qualitative techniques, falls clearly into the category of moral exhortation that many of his peers tried so strenuously to supercede— not because his views were disrespected—but because they lacked substantiation.

In seeming contrast, Riis had a freeing imagination: he worked to jar a well-to-do population out of complacency: walls should be torn down, tenements razed, the sky admitted to view. His imagination drove him to let loose, to open up, as if sunlight and air were themselves solutions to political and economic problems. His most valued achievement was also an emblematic one: the destruction of a foul area of the East Side—Mulberry Bend—and the establishment of a public park in its stead. Space itself signalled the defeat of the landlord class. As if in direct rebuke to Brace, Riis reversed his predecessor's system of causality: it was not character but environment that threatened social stability. And since he regarded environment as malleable, he attacked those whose financial interests prohibited civic improvement. "The 'dangerous' classes of New York long ago compelled recognition," he said. "They are dangerous less because of the criminal ignorance of those who are not of their kind. The danger to society comes not from the poverty of the tenements but from the ill-spent wealth that reared them."[65] In short, the "dangerous" classes were not the poor, but those who held them captive. Riis wrote to dispel the notion that Brace wanted to propagate: that poverty was dangerous. He wanted instead to speak on behalf of "thousands misjudged by a happier world, deemed vicious because they are human and unfortunate."[66] This is a feature of Riis's thinking that will be essential to remember when we come to evaluate his photographs more fully; for though his images seem to feature downtrodden people, they are more directly intended to be read as "traces" of an absent but maliciously negligent landlord. Riis's imagination was essentially theistic, and he thought of the tenements as features of a dark, cosmic drama; they were evidence of sinful deviance from the clean air and sun of the original creation; and his own role, consequently, was that of the persecuted evangelist, the harbinger of light: "I was Paul," he said, "preaching salvation to the Jews."[67]

I mention this characteristic of Riis's intellect at this point partly as a warning and partly as an attempt to temper what follows, for Riis refused to write a book of moral exhortation of the type Brace had done; and he would not have described his work as I have done—as an imaginative morality play. In his own eyes *How the Other Half Lives* was an objective document which was properly buttressed with charts, statistics, and facts.

Riis's own talent, gained from years of journalism for *The Tribune,* was to isolate the poignant vignette: "Right here, in this tenement on the east side of the street, they found little Antonia Candia, victim of fiendish cruelty, 'covered,' says the account found in the records of the Society for the Prevention of Cruelty to Children, 'with sores, and her hair matted with dried blood.'" But he refused to let sentiment suffice. "In this block," he continued, "between Bayard, Park, Mulberry, and Baxter Streets, the late Tenement House Commission counted 155 deaths of children in a specimen year (1882). Their percentage of the total mortality of the block is 58.28, while for the whole city the proportion was only 46.20."[68] This combination of anecdote and quantitative generalization is one of Riis's typical narrative strategies. Dr. Roger S. Tracy, Registrar of Vital Statistics of the New York Board of Health, provided Riis with most of the statistical information that he used in *How the Other Half Lives,* and by including it Riis considered that he had entered into the cultural debate about proverty as a positivist with undisputable evidence. "These figures speak for themselves," he repeatedly asserted. He used them not only because of the anticipated antagonism of those whose financial position he threatened, but also because of disputes about appropriate methodology that existed within the social scientific community itself.

Whatever Riis's self-justification, his approach to writing about the "other half" was full of confusion: he had a Christian mission, he claimed to write objectively, to deal in facts, and he based his authority on his own experience in America, on an entry into New York City which had been full of the same hope and rebuff of circumstances that he saw repeated in the lives of other new immigrants:

> I joined the great army of tramps, wandering about the streets in the day time with the aim of somehow stilling the hunger that gnawed at my vitals, and fighting at night with vagrant curs or outcasts as miserable as myself for the protection of some sheltering ash-bin or doorway. It was under such auspices that I made the acquaintance of Mulberry Bend, the Five Points, and the rest of the slum with which there was in the years to come to be a reckoning. For half a lifetime afterward they were my haunts by day and by night, as a police

reporter, and, I can fairly lay claim, it seems to me, to a personal knowledge of the evil I attacked.[69]

Because they are incompatible, these three postures reveal the naïveté of Riis's empiricism: he used facts only as secondary examples or confirmation of positions already held. Even had he worked impartially, gathering information that cumulatively pointed to "self-evident" conclusions, those conclusions, those ends, were also means—they formed part of a larger argument about the damaging behavior of landlords and the subsequent need for community involvement in matters of adequate housing. This argument, in turn, rested not on science in any form but on a Christian world view in which greed or disregard for the sanctity of individual and family life was an affront to God. Riis was an evangelist; his aim was intervention; he never intended for dispassionate observations to leave things as they were.

The reasons for calling attention to these conceptual confusions are fairly simple: they show, more clearly than much of the social science writing of the time shows, the nature of the interest that can shape social inquiry and preclude objectivity, even when the author's self-understanding is that he is unbiased. They account for the ambiguity of the speaking voice in the text, and they allow us to see that Riis's use of statistics about overcrowding, disease, lack of sanitation—his "facts"—were always part of a rhetorical strategy to persuade others to see the world as he did and to intervene in changing it. *How the Other Half Lives* is a book that supports, rather than discredits, Edward Youmans's criticism of American social scientists—that they "deliberately attempt[ed] to deceive the public by cloaking their reformist schemes with the dignity of science."

In saying this, Riis's confusion is nonetheless one of the most important aspects of his book; that is, his problem in identifying with either one "half" or the other in a culture radically divided into an assimilated majority and an outcast other is what calls for our most careful attention: because the voice of experience, the voice that Riis used naturally when he remembered himself as a poor tramp, was the voice of the subject of social inquiry. The voice of the fact-gatherer, the social investigator, was the voice of authority, the voice through which ideas about the very structure of society and the nature of its population were authenticated and then communicated to the public as true. His own remarkable social mobility, his rapid rise from destitution to middle-class security gave him, in a sense, two languages. Riis knew intuitively that these voices in writing—and perspectives in photography—stood in a profoundly important relation to each other. He recognized that they each embodied valid, though conflicting, authority even though the positivistic methodology of the emerging social sciences told him to invalidate and silence the voice of subjective memory and experience.

This double awareness is present throughout *How the Other Half Lives*. On one hand, Riis is conscious of having special access to places because of his old familiarity with street and tenement life; on the other, he is conscious of being abrasively excluded on those occasions when he was understood to represent the interests of formal authority. His familiarity with "the poor" is demonstrated in the text by recreated dialogues between himself and the people he meets and subsequently describes. These vignettes typically follow the pattern established by his "facts": they are further proof of the good character of the immigrants who are thwarted by wretched living conditions:

> "How many people sleep here?" The woman with the red bandana shakes her head sullenly, but the bare-legged girl with the bright face counts on her fingers—five, six! "Six, sir!" Six grown people and five children.
>
> "Only five," she says with a smile, swathing the little one on her lap in its cruel bandage. There is another in the cradle. And how much the rent?
>
> "Nine and a half, and Please, sir! he won't put the paper on."
> "He" is the landlord. The "paper" hangs in musty shreds on the wall.[70]

But, unlike facts, conversations must be elicited in person and their interest to us here, far more than their manifest content, lies in the revelation of the relation of speaker to subject; Riis has gained the trust of the child, but not of the parent, for whom, presumably, he was indistinguishable from other authorities. Indeed this assessment of his reception is reconfirmed in another situation with a street gang: "[H]aving my camera along, [I] offered to 'take' them, They were not old and wary enough to be shy of the photographer, whose acquaintance they usually first make in handcuffs and the grip of a policeman."[71] In each case, as Riis realized, ease of access, whether to social knowledge or visual information through the camera, was dependent upon Riis's perceived relation to authority, to those with the power to intervene in the subjects' lives. This is an essential connection to understand, for the kinds of dialogue Riis engages in correspond to the types of photographs he is able to take; they serve as a convenient index of—or way to categorize—his visual images. All of this information, in turn, tells us what Riis thought he recorded as he made his rounds with camera and flash powder; it clarified his relation to the world of appearance, to the evidence available in the visible world.

Returning first to the issue of dialogue, we can see, even from these two sparse examples, that Riis imagines communication to occur in two opposing modes: the informative mode derives from innocence; the wary and withholding mode is a posture assumed from the self-conscious experience of powerlessness. A similar polarity characterizes his photographs. To the extent that he can acquire trust, he can take those pictures in which families or individuals sit in obviously arranged positions: some families look fully into the camera wearing their good clothing; some continue to engage in their normal work; some implicitly make a point about specific aspects of life—a typical degree of literacy, a typical meal. These types of images imply some degree of structural equality between the photographer and his subjects; they speak to us in Riis's "first voice," the voice of empathy and of remembered and shared experience. They are, consequently, the least problematic and in some ways the most eloquent of Riis's pictures.

The more difficult images to assess are those "spoken" in Riis's "second voice," those that tell us, through stunned or defiant gazes, that Riis's presence in the midst of his subjects created difficulty and evoked resistance. Of the Jewish response to health officials, Riis at one time observed, "Their first instinct [was] to hide their sick lest the authorities carry them off to the hospital to be slaughtered, as they firmly believe."[72] But he also recognized that his camera elicited similar fear or stealth. One need not look far to understand why. In his autobiography, he remembered the origins of his photographic expeditions as the bringing of light into darkness; he conflated light and enlightenment, flashbulb and insight; and he treated his subjects with messianic zeal as if they could not possibly understand the darkness in which they lived.

> One morning, scanning my newspaper at the breakfast table, I put it down with an outcry that startled my wife, sitting opposite. There it was, the thing I had been looking for all those years: a way had been discovered to take pictures by flashlight. The darkest corner might be photographed that way. Within a fortnight a raiding party composed of Dr. Henry G. Piffard and Richard Hue Lawrence, two distinguished amateurs, Dr. Nagel and myself, and sometimes a policeman or two, invaded the East Side by night, bent on letting in the light where it was so much needed.[73]

> It is not too much to say that our party carried terror wherever it went. The flashlight of those days was contained in cartridges fired from a revolver. The spectacle of half a dozen strange men invading

a house in the midnight hour armed with big pistols which they shot off recklessly was hardly reassuring, however sugary our speech, and it was not to wonder at if the tenants bolted through windows and down fire escapes whenever we went. Months after I found the recollection of our visits hanging over a Stantion Street block like a nightmare.[74]

One of his most arresting pictures provides evidence of such terror and nightmare. It is entitled "Lodgers in a Crowded Bayard Street Tenement-'Five Cents a Spot'"; and it is intended to represent conditions of squalor and degradation that characterized "The Bend," a district between Broadway and the Bowery and Canal and Chatham Streets. Riis describes the circumstances that led to the image:

The doors are opened unwillingly enough—but the order means business, and the tenant knows it even if he understands no word of English—upon such scenes as the one presented in the picture. It was photographed by flashlight on just such a visit. In a room not thirteen feet either way slept twelve men and women, two or three in bunks set in a sort of alcove, the rest on the floor. A kerosene lamp burned dimly in the fearful atmosphere, probably to guide other and later arrivals to their "beds," for it was only just past midnight. A baby's fretful wail came from an adjoining hall room, where, in the semi-darkness, three recumbent figures could be made out. The "apartment" was one of three in two adjoining buildings we had found, within half an hour, similarly crowded. Most of the men were lodgers, who slept there for five cents a spot.[75]

When Riis worked in these ways, his motives could sometimes not be discerned. In the eyes of his subjects, his methods of gathering images placed him not in opposition to the police—where Riis would have placed himself—but in alliance with them. The ramifications of this are crucial, for they show us a benevolent photographer with a private, antiauthoritarian purpose reduplicating rather than reversing the movements of official power. How could the Bayard Street tenants distinguish Riis from the police or health or truant officers whose rules about illegal overcrowding threatened whatever thin security numbers could bring them? Riis could only be understood by them as someone gathering information for uses that remained hidden from their view.

And, in fact, this was not entirely a misunderstanding, for Riis always considered that he spoke on behalf of the poor whether he photographed them in the dignity of their small, created orders or in the misery of haphazard and cruel circumstances. The voice of empathy was not, nor could it be, the voice of poverty itself. Riis was no longer poor, and in fact it was his own securely established suburban house-

hold that both defined his social aims and gave him the peace of mind
to advocate them: it was distress remembered and escaped that allowed
Riis to write and photograph as he did. And it was to others who were
also economically secure that Riis addressed himself. Peter Hales has
suggested that Riis's photographs are encoded "in the symbolic language of late nineteenth century American middle-class Victorianism,
invoking that culture's most cherished ideals—womanhood, the home,
the child, work, the separation of the sexes and races, cleanliness, light,
air, health, education, religion, and above all, order."[76]

I think that this is a significant way to characterize Riis's images,
for it suggests that the most powerful dialogue in *How the Other Half
Lives* is not between Riis and the subjects of his inquiry, but between
Riis and his middle-class viewers. It is not an accident that the vocabulary of the photographs is provided by the secure rather than the
nether world, that most of these photographs achieve meaning by
implicit reference to their wholesome, absent counterparts: the slum's
dirt evoking suburban cleanliness; its crowding   light, air, and privacy,
its indolence—labor; its street urchins—the well-cared-for and treasured child.

In his own mind, Riis's aim was a simple one: to make the world
he has made visible—with its affront to established values—a catalyst
for action. This relationship points, in turn, to Riis's assumptions about
the nature of the social world taken as a whole: that it is, and will
continue to be, a place of actors and those acted upon—not because of
the slum landlord, whom he characterized as the chief villain of the
tenement world, but because of the conceptual structure of his own
methodology as a fact-gatherer. That is, Riis's very way of assuming
the role of social science observer—and by extension his way of using
the camera in this project—extended rather than diminished the dom-
inant-submissive (upper and "other" half) model of social organization
that he ostensibly attacked. The role of the immigrants was to ask for
help. The role of the photographer was to amplify their pleas and to
ensure them an audience capable of effective political action.

The questions that remain unanswered by *How the Other Half Lives*
are manifold and instructive: what kind of self-understanding lay
behind the signs of urban hardship Riis recorded? What capacity did
his subjects have for moral relationship or for communicative interaction with the self-appointed agents of "culture" who measured,
described, and photographed them? What solutions did they envision
for their own plight?

Charles Loring Brace, looking back on his own career as a social
reformer, called attention to one of the central problems obscured by

Riis's evangelical zeal and naïve positivism: in one of the journals he kept during twenty years of work in the New York slums, he singled out the story of a young prostitute who had later reformed:

> After a short waiting, the girl was brought in [to the Tombs]—a German girl, apparently about fourteen, very thinly but neatly dressed, of slight figure, and a face intelligent and old for her years, the eyes passionate and shrewd. I give details because the conversation which followed was remarkable.
>
> The poor feel, but they can seldom speak. The story she told, with a wonderful eloquence, thrilled to all our hearts; it seemed to us, then, like the first articulate voice from the great poor class of the city.[77]

"It seemed to us, then, like the first articulate voice from the great poor class of the city." Riis, for all of the reasons I have tried to enumerate, did not recognize the value of letting poverty reveal itself to others, of allowing the self-understanding of the poor to play a role in their own representation and the policies ensuing from it. Theodor Adorno has warned us about the results of cultural strategies like Riis's: "If we attempt to place subject and object on opposite sides [in the other half], we shall fail to comprehend either one or the other."[78] Riis, in speaking for the immigrant population of the Lower East Side of New York City, left them as silent, as voiceless, and as subject to control as his more purposefully manipulative colleagues. The value of his book, and its interest to us in the study of the photography and literature of social investigation in America is that Riis's confused identity, and the ambiguity of photographic perspective ensuing from it, points to the possibility of the subjects' of inquiry having their own voice, their own image of themselves, and the motive to engage in reciprocal communication with others.

VII

Dreiser struggled with finding an authentic perspective for most of his writing career, but the issues of seeing and class identity were joined most forcefully and most personally in The "Genius"—the novel which eventually incorporated the material gathered from his 1902–1904 period of crisis and manual labor. Indeed, the book can be read as a demonstration of the social and political implications implicit in the two modes of seeing represented to him by Stieglitz and Riis. It enacts the dilemma of an artist who moves from an art very similar to Stieglitz's to a recognition of all that lies outside of that art and remains unrepresented by it. It is Dreiser's explanation of outgrowing the photographic picturesque and assuming an aesthetic identity that more clearly admits the damage of his own experience within the unnamed stratifications of American cultural life.

The novel is cumbersome, and its drama lies in the nuances of psychological trauma rather than in the confrontation of the protagonist, Eugene Witla, with the larger social world. Nonetheless, the text documents the protagonist's movement toward the integration of personal perspective and representational modes: while Witla begins his artistic career with a brilliance of style, it is not, finally, an expression of his own experience of life; it is borrowed from a chic, urban crowd whose preferences the young man tries initially to accommodate and adopt.

Dreiser represents Witla's early career in New York City as the heeding of temperament, the following of the same kind of unexamined fascination with spectacles and street scenes that had characterized the first artistic sketches of his own move to the city:

> Eugene had covered almost every phase of what might be called the dramatic spectacle in the public life of the city and much that did not appear dramatic until he touched it—the empty canyon of roadway at three o'clock in the morning; a long line of giant milk wagons . . . the bread line. . . . Everything he touched seemed to have romance and beauty, and yet it was real and mostly grim and shabby.[79]

These pictures appeal to the critics and gallery dealers in the novel; they establish Witla as a young, talented artist. Their interest as a cipher in the symbolic play of Dreiser's imagination over the direction of his own career is that the canvases, almost without exception, constitute a descriptive catalogue of Stieglitz's New York City photographs: "One of Fifth Avenue in a snow storm, the battered, shabby bus pulled by a team of lean, unkempt, boney horses. . . . He liked the delineation of swirling, wind-driven snow. The emptiness of this thoroughfare, usually so crowded, the buttoned, huddled, hunched, withdrawn look of those who traveled it, the exceptional details of piles of snow sifted onto the window sills and ledges" (219) is, of course, Stieglitz's "Winter on Fifth Avenue, 1892–1893." "[T]he three engines entering the great freight yard abreast, the smoke of the engines towering straight up like tall whitish-grey plumes, in the damp, cold air. . . . You could feel the cold, wet drizzle on the soppy tracks, the weariness of 'throwing switches'" (223) could be either Stieglitz's "Hand of Man, 1902" or "In the New York Central Yards, 1903." "The steaming tug coming up the East River in the dark hauling two great freight barges" is "The City of Ambition, New York, 1910." "Greely Square in a drizzling rain . . .

by some mystery of.... art had caught the exact texture of seeping water on gray stones in the glare of various electric lights, those in cabs, those in cable cars, those in shop windows" (219) could be drawn either from "Reflections, Night, New York, 1897" or "Wet Day on the Boulevard, Paris, 1894."

The point is not that Witla is a fictionalized Stieglitz, but that the novel dramatizes Dreiser's view of the adequacy of such art—not to the critics who, also like Stieglitz, approve of its realism and hang it on gallery walls—but to the artist himself. The collective work is described this way: "Why this thing fairly shouted its facts. It seemed to say: 'I'm dirty, I am commonplace, I am grim, I am shabby, but I am life.' And there was no apologizing for anything in it, no glossing anything over. Bang! Smash! Crack! came the facts one after another, with a better, brutal insistence on their so-ness" (222–23). But such encomiums occur within a novel about colossal failure and self-redefinition; the Witla who plunges into the New York art world with these images crashes out of it in an equally dramatic way. By his late twenties, he is in the midst of a deep personal crisis which soon leads to a full breakdown.

In *Dreiser's Novels,* Donald Pizer makes the argument that *The "Genius"* is not primarily a portrait of the artist but a portrait of a grotesquely miserable marriage, and since he sees the mismating of Witla and his wife as the central theme of the book, he considers the marriage to be the cause of Witla's failure. Read in this way, *The "Genius"* is a heavy, plodding variation on the vampire theme, worked so well by Henry James, where a shrewish woman drains the artist of his creative powers. Had he not married this woman, he would not have been diverted from the use of his greatest gifts. But to argue in this manner is to ignore the nature of Witla's art and to evade questions about the function of art in the life of the artist. For, read in another way, Witla's dilemma comes not simply from a difficult woman, but from the radical dissonance between the pain and poverty of his own private life experiences, and the romanticism about the social order which he tries to maintain on his canvases: it is only to the privileged onlooker that the "drama" of what is ugly can be maintained.

This is a connection that Witla does not make quickly or easily. Initially he is simply awash in pain from living with a woman he does not love and comes increasingly to despise. In that these events have intrinsic interest as a candid portrait of marriage, Pizer's assessment of the novel is a good one. However, this inchoate pain leads to other pain with a more discernable shape to it: Witla handles his mental breakdown, much as Dreiser did himself, by leaving home and working with his hands as a day laborer. The two experiences of loving and working have deprivation and powerlessness in common, and Dreiser frames both situations in the language of exclusion: they show Witla what he is outside. In love, the matter is rather simply expressed: "She could not understand the agony of a soul that had never really tasted supreme bliss in love, and had wanted, however foolishly, the accessories of wealth and had never had them" (564). This unsatisfied desire is handled rather simply by a series of clandestine affairs which succeed, whatever their erotic or spiritual benefits, in heightening the jealousy and hysteria of his wife. Witla's pain may be momentarily displaced, but it is not fundamentally assuaged. Nor does his behavior bring either self-reflection or control.

In his factory work, Witla gains more insight—and for an interesting reason. Whereas marriage and art had remained irrevocably dissociated to Witla, factory work and art acquire a curious congruity: by joining the factory in Speonk, New York, he has, in effect, entered the frame of one of his own industrial pictures to participate in what he had only witnessed before. To be sure, he brings his pencil with him, draws as he did before, and seeks the old artistic immunity, but the work itself changes him. It robs him of the "aesthetics of ugliness" and it returns him to himself as a worker, as a man with a sense of the importance of his subject and with an understanding of the relation of the private pain of experience and the communal pain of the victims of an unseen social hierarchy.

Where his marriage had left him with only an indiscriminate sense of lack, the deprivations of manual labor lead him to understand that he lives in a world that engineers structural inequalities. Formerly the hierarchies of power in the universe were inexplicable to him—all out of order; but here, "he saw by degrees." Witla initially begins his work at the railroad carpenter center as a distrusted outsider, as someone connected with authority. "The foreman went on his way, thinking. If Eugene could [draw] that, why was he here? It must be his run down condition, sure enough. And he must be the friend of someone high in authority.... He could not figure out just why he was here—a spy possibly" (317). In the eyes of his fellow workers, his talent gives him a special status; but the main point is that his work experience effects a change of consciousness so that he ceases to idealize the industrial world. The shop is "a veritable song of labor" until he is forced to sing the weighty song himself. "Eugene straightened up and the rough post balanced itself evenly but crushingly on his shoulder. It appeared to grind his muscles and his back and legs ached instantly. He started bravely forward straining to appear at ease but within fifty feet he was

suffering agony" (309). This personal disenchantment is soon generalized to his vision of the entire industrial scene:

> At the same time, now that the fall was coming on, he was growing
> weary of the shop at Speonk, for the gray days and slight chill which
> settled upon the earth at times caused the shop windows to be closed
> and robbed the yard of the air of romance which had characterized it
> when he first came there. . . . He was beginning now to see also that
> [Big John and Joseph Mews and Malachi Dempsey] were nothing but
> plain working men (353).

Finally Witla comes to identify the artist as another kind of worker
who is subject to the same system of social stratification as the Speonk
workers; and it is precisely this recognition that led Dreiser to say of
himself in *Newspaper Days*, "[T]he sad state of the poor workingman
was a constant thought with me, but nearly always I was the greatest
and poorest and most deserving of all workingmen."[80] The consequences of this insight are almost immediate in Witla's life. He returns
to New York, not to paint more romanticized urban, industrial pictures, but to practice commercial art, to join the business world of
which he has seen himself to be a rather lowly part. Witla's career in
magazine advertising has been seen by characters in the novel and by
critics of the novel as a detour, as another misapplication of "purer"
talents, but in fact it is the logical consequence of his experience as a
day laborer. It is a natural and honest response to the identification of
himself as part of an economic hierarchy, and it has occurred almost
in spite of any conscious search for more authenticity. At first intended
to be a cure for overwrought nerves, looked at as a "picturesque" addition to his experience, the Speonk carpenter shop restores an essential
bond between the artist and the subject of his art, and shows Witla that
his debilitation was inextricably connected not with a single woman—
although that would constitute an idiosyncratic explanation—but with
a more fundamental dissonance, a failure of connection, between the
seer and the seen, between his situation as an artist, as a worker in art,
and the larger cultural situation of which he is a part.

The *"Genius"* blusters off into one of Dreiser's unsatisfactory end-
ings: Witla connives in business and fails; he plays for grand passion
and loses at that too. He finds solace in the renunciations of Christian
Science, and Dreiser resolves all of these cataclysmic events by reference to "the unknowable." What Witla paints or what he thinks of his
art remains unarticulated. Nonetheless, the text tells us that for Dreiser,
observation that caresses the surface and remains captivated by the
beauty of evanescent details henceforth would be considered a cultural

strategy that evaded the deeper structural issues that accounted for those details. An art that conflated beauty with an inherently grim subject, that erased pain in the service of a picturesque spectacle, would, from this point, be an inadequate art for Dreiser. And although the surface, the appearance of things, would remain of strategic importance to him as a way to free the interpretation of things from their embeddedness in traditional ideological explanations, his own search from this point would, like Jacob Riis's, be to see what lay underneath.

If *The "Genius"* commemorated Dreiser's refusal of supposedly disinterested vision as an appropriate vehicle for his own art, it did not bring to a close his fascination with the aesthetic disengagement and manipulation that characterized other aspects of American culture. Indeed it is possible to consider one of Dreiser's major achievements to be his association of calculated observation with the business community rather than with the artistic world. Although he himself would no longer rest contentedly as an impartial observer, delighting in the picturesque, he would never lose his fascination with those men of financial genius who managed the lights and shadows of industry in much the same way that Hawthorne had managed the lights and shadows of his own dusky fictions. In fact, Dreiser came to see artists like Hawthorne, James, and Stieglitz as the precursors of the robber barons of the Gilded Age and, conversely, to understand his own perspective as belonging more authentically to "the other half" whose lives he had briefly entered as a convalescing day laborer. That is, Dreiser's choice of representational modes became increasingly entwined with an understanding of the financial and social implications of the artist's status as a worker, and his choice of subject became increasingly influenced by the kinds of cultural manipulations he could understand because he had once engaged in a similar disregard for his real relation to the scene before him. In Dreiser's work, Eugene Witla's aesthetic heir is not another, more mature artist, but a financier, a giant of the corporate world, a builder of industrial landscapes rather than of "vital," "raw" canvases. Dreiser reassigns his observational posture to an industrialist, and then uses his own prior knowledge of the mechanics of such manipulation to expose or unmask the dynamics of power that would otherwise remain undisclosed to his readers.

If he began his artistic life in the shadow of Alfred Stieglitz, he entered into full maturity with the allegiances of Jacob Riis and with the decision to use art to deal, not with the arrangement of surfaces, but with the hidden structures of the cultural scene. To expose became an increasingly important goal for Dreiser; to avoid exposure became the obsession of his characters. It is in light of these complimentary tendencies of scrutiny and avoidance, of surveillance and escape, that his Cowperwood trilogy, the great saga of commercial triumph and collapse, can be read; for Dreiser sets up scenes of private and commercial achievement only to undermine them by exposing their duplicitous foundations. Dreiser seems to work almost ritualistically, repeating paradigmatic structural situations where duplicity is undermined not by superior force, but by superior knowledge. In these novels, then, observation has the effects of power. In the same way that Jacob Riis considered that the camera could tear down tenement walls by the revelation of what lay behind them, Dreiser considered that art could act in the world by unveiling the questionable premises of apparent success.

We see this in the Cowperwood trilogy where Dreiser is interested in both the career of a strong-willed industrial magnate and in his defeat. Dreiser's model for the trilogy (*The Financier*, 1912; *The Titan*, 1914; *The Stoic*, 1947) was an actual Philadelphia business man, Charles T. Yerkes, whose life he researched meticulously and transcribed with a sometimes irritating scrupulousness.[81] The young Cowperwood is shown to be shrewd, dynamic, and aware that business is as much a matter of managing human psychology as it is of manipulating stocks. The opening scene of *The Financier* in which the child, Frank, watches a lobster devour a squid is usually taken to be emblematic of the entire structure of the boy's later life, a life characterized by brutal egoism and an unrestrained will to power: by the time Cowperwood is thirty-four and serving a prison term for larceny and embezzlement of funds from the Philadelphia treasury, he has already achieved (and lost) the wealth and social position that others acquire, if at all, during an entire lifetime. But as much as this is a novel about the acquisition of power, it is also a novel about the accumulation of information; and in Dreiser's eyes, the two are inextricably related: the lesson about brute force that the young Cowperwood learns is followed quickly by another. Cowperwood's mature life is characterized by knowing—when to borrow illicit money, what car line franchises to buy, which politicians to favor—and by furtiveness. "There was another man his father talked about—one Francis J. Grund, a famous newspaper correspondent and lobbyist at Washington, who possessed the faculty of unearthing secrets of every kind."[82] The secret of success is secretiveness: information is useful to the extent that it is clandestine, exclusive, inaccessible to others. What gives Cowperwood's career added interest is Dreiser's tendency to render his character's private life in almost exactly analogous terms: the secret love affair is more valuable than the open marriage; and in both cases—the business arrangement and the love affair—danger is the inevitable companion of illicit advantage. Cowperwood moves easily among these tensions, maintaining an inscrutable mien, manipulating women and money, until the Chicago fire of 1871 shakes the Philadelphia business establishment.

What is important about this disaster in Dreiser's world of subterranean maneuvers is simply that it forces disclosure. Cowperwood is caught with $500,000 of the city's money invested in personal ventures, and while such underhand dealings characterize the entire commercial/political world of the novel ("If only the great financial and political giants would for once accurately reveal the details of their lives"), Cowperwood fails because he alone among all the others can no longer conceal his activities: if power is acquired through secrecy, it is lost through exposure, and indeed, the end of the novel is a kaleidoscope of moves by those in exactly analogous positions who maneuver to cover themselves. As Cowperwood's lawyer characterizes the situation to a jury, "The entire uproar sprang solely from the fear of Mr. Stener at this juncture, the fear of the politicians at this juncture, of public expose. No city treasurer had ever been exposed before. It was a new thing to face exposure."[83] As Cowperwood himself explains to his mistress, Aileen, "If this should end in exposure, it would be quite bad for you and me." In Cowperwood's life situation, depending as it does upon secrecy, to become visible is to become vulnerable to the control of others; he finally is stripped of business, family, and temporarily of his lover, and if Dreiser resurrects him in The Titan and The Stoic, it is to explore further this hierarchy of power in which observation itself can conquer even the most invulnerable of social engineers.

The further interest of the sequels lies in Dreiser's association of Cowperwood's manipulations with the activities of the artist—or at least with artists who have certain kinds of aesthetic propensities. It is here that Dreiser's rejection of Stieglitz and pictorial expressionism is most fully articulated and shown to have relevance to American culture as a whole. For the decisions and aesthetic choices of a single avant-garde photographer have one order of importance, but they acquire a much more substantial status when they characterize a part of the American corporate world as well. Cowperwood is, in Dreiser's eyes, the unseen manipulator, using people, resources, and land for compositions every bit as aesthetic as Stieglitz's "Hand of Man" or "The Steerage." This is a characteristic that Cowperwood sees in himself. He believes "that men like himself were sent into the world to better perfect its mechanism and habitable order," and it is also a trait that others come to recognize. "She [Stephanie Platow] conceived of him as a very great artist in his realm rather than as a businessman." And as Cowperwood plans his assault on the Chicago street railway system—a project to restore the fortune and prestige he lost in Philadelphia—he watches the city as one might look at a painting, or frame a street scene for a photograph. His mode of envisioning it is analogous to Eugene Witla's; the scene is dirty, odorous, compact, with "picturesque tugs in the foreground below." Finally Dreiser makes the aesthetic nature of Cowperwood's impulse to build and control explicit by remarking, "He forever busied himself with various aspects of the scene quite as a poet might have concerned himself with rocks and rills." In these comments Dreiser conflates all of the arts, seeing the poet, the painter, and the photographer as unseen manipulators, using the material world for projects of egoistic importance, using resources without concern for the effect such use might have on the subjects of the vision. Cowperwood's goal is similarly to control and to own, to appropriate the physical world, to subordinate it to his own will and to accomplish this without being seen. He is, in this respect, not only the heir of Stieglitz (as Dreiser understood him) but also of Hawthorne and James, and this imaginative lineage is quite direct: where Hawthorne described his artistic activities as those of a stage manager manipulating his "atmospherical medium as to bring out or mellow the lights and deepen and enrich the shadows of the actual world," Cowperwood sees himself "in the position of a man who contrives a show, a masquerade, but is not permitted to announce himself as showman." He is busy disguising the origin or presence of his own power.

Where James experienced repeated anguish over "the expense of vision" in his art, Cowperwood bears the animosity of his subjects through a succession of financial and personal disasters. When foiled once again by public opinion in Chicago, brought low by a people enraged at his underhand dealings, he asks why he repeatedly arouses such hatred and decides that the issue is not morality, but the invisibility of his power, his ability to control "without standing out fully and clearly in the sight of men." Cowperwood's distinction, of course, lies in his attitude toward his own posture in life. Where his precursors experienced the full burden of a dilemma, where they understood themselves to be caught in a web of antithetical creative and moral demands, Cowperwood sees nothing problematic about his own "art" of management. Life remains for him the grotesquely simplistic set of circumstances represented to him in youth by the devouring lobster in the tank. It is a matter of unadulterated power relationships. Although he would prefer to conceal the mechanics of power and to work without being seen at it, he never acquires more self-reflection or more clarity.

What is essential to recognize about the novels that tell this story of ruthless and unreflective strength is the rift between the artist who invents the text and the emblematic power relations set up within it. For Cowperwood does not reenact the dilemma of Dreiser himself in the way that Holgrave became a cipher for Hawthorne or the Princess Casamassima a cipher for James's creative impasse. Instead he represents a set of rejections and is looked at by his creator from a very different vantage point. Although Dreiser understands Cowperwood's perspective, he writes about it to expose it—to call attention to it and to condemn it as a personal philosophy that is, finally, inadequate to life.

In the novels the viewpoint that more nearly represents Dreiser's own allegiances is an amorphous one, present to us in the form of a faceless public whom Dreiser, at an earlier point in his career, would have spurned and feared. But with his own identity as an artist more firmly established, the need to distinguish himself from the undistinguished mob diminished, so he understood himself to have roots among the common people most damaged by the manipulation of high finance. He came to see also that for those whose lives are shaped by the large-scale use of capital, the use of capital is usually invisible. To them its existence and its power become known, if at all, through a symbolic figure, a villain whose lifestyle evokes both loathing and fascination.

Neither Riis nor Dreiser was a logician; they achieved an understanding that was in some ways enlightened yet retained other attitudes that were analogous to those held by the "other half" whose interests they came eventually to espouse in their art. Both understood that American economic society was radically polarized, and both villainized and blamed a small class of men for the problems they saw: Cowperwood was for Theodore Dreiser what the landlord was for Jacob Riis. These members of the ruling elite were, in Dreiser's words, "Titans—who, without heart or soul, and without any understanding of or sympathy with the condition of the rank and file, were setting forth to enchain and enslave them."

In the thirties, Dreiser began to turn these inchoate sympathies into a more systematic ideological position, and with the publication of *Tragic America* (1931) and the editing of *Harlan Miners Speak* (1932), he identified himself as a socialist. However, this political philosophy was not known to him when he wrote *The Financier* and *The Titan*. His struggle in the Cowperwood books was of a different nature: he had rejected Stieglitz's model of the artist and saw himself to be very much like Jacob Riis. That they probably did not know each other and might have disliked each other has no bearing on the similarities with which they conceived their creative mission: for Dreiser, the writer, and Riis, the photographer, were engaged in a self-initiated struggle which pitted real political strength which they deplored against the power of observation itself. Their aim, with both the camera and the pen, was to put an end to secrecy, to reveal the corruption shielded by impenetrable walls, and to use disclosure itself as an alternative and hopefully superior means of control for those who could come to name and thus potentially act on the unseen forces that shaped their lives.

Alfred Stieglitz. The City of Ambition. New York, 1910 *(IMP/GEH)*

Jacob Riis. Pietro Learning To Write: Jersey Street *(Museum of the City of New York)*

Jacob Riis. Lodgers in a Crowded Bayard Street Tenement: Five Cents a Spot *(Museum of the City of New York)*

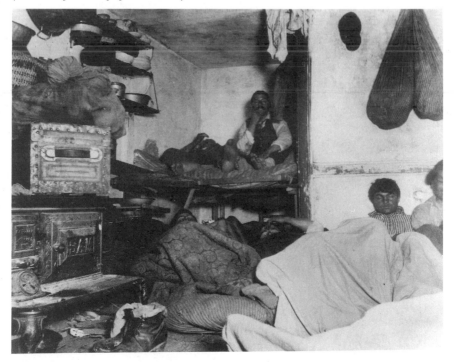

Joris Ivens. Film still from *Borinage (IMP/GEH Still Collection)*

4

John Dos Passos
and the Soviet Cinema
Separate Frames of Truth

These frames must be thematically organized so that
the whole is also truth. —Dziga Vertov

I

IN *Manhattan Transfer,* John Dos Passos's first novel about New York,
one of the characters is haunted by the dream of a skyscraper. It is
a "grooved building, jutting up with uncountable bright windows
falling onto him," and he cannot find its entrance: "Please mister
where's the door to this building?" he asks in the midst of frenetic
circling. The door, of course, is in front of him; it is only in imagination that he is shut out; but it is emblematic of Dos Passos's fictional
world that characters should feel closed away from their lives in the
midst of living them. Possessed of the keys to the city, able to enter
affluent and glamorous careers, they are nonetheless as estranged in
their own beds as Dreiser's characters had been in the open and undefended streets:

> The faraway sounds of sirens from the river gave him gooseflesh.
> From the street he heard footsteps, the sound of men and women's
> voices, low youthful laughs of people going home two by two....
> There came on the air through the window a sourness of garbage, a
> smell of burnt gasoline and traffic and dusty pavements, a huddled

stuffiness of pigeonhole rooms where men and women's bodies writhed alone tortured by the night and the young summer.[1]

To the extent that Dos Passos makes this anomie the subject of reflection within the novel, he does so through a young newspaper reporter, who dreams of his own exclusion from life and who balks at the "pigeonhole" nature of the lives he and his friends lead in Manhattan. Piled into tenements, shoved into subways, elbowing each other on the streets, they remain solitary, framed by their own subjectivity, contiguous but untouched. For Dos Passos, the still photograph expressed their condition and became a model—though a problematic one—for his narrative art. His struggle in the early days of his career was to discover a novelistic device for rendering the lives of people who remained exclusively concerned with form; he needed to find an appropriate way to embody the character of those who were, in his judgment, brittle, lacking in interiority, and concerned only with the "beauty" of their own images. One might think of the characters in Dos Passos's early fiction as a collection of Stieglitz photographs, a set of living testaments to the hazards and limitations of excessive concern for the aesthetic surface. Like Dreiser before him, Dos Passos criticized such formalism as a tendency in life and as a practice of art, but unlike his predecessor, who remained obsessed with secrecy and with surfaces as screens for something unexpressed—the subterranean manipulations of high finance or clandestine sex—Dos Passos grew to be concerned primarily with the photograph as an austerely circumscribed view, a fragment.

Where Dreiser abandoned his fascination with beautiful surfaces and, like Riis, came to see the role of the artist as an exposer, as a visionary making visible that which special interest groups were concerned to hide, Dos Passos at first concentrated on the artist's negative function, his exclusion from the field of vision, his limited range of vision, and yet the extraordinary importance accorded to vision in a culture that seemed unable either to feel or express depth or to find connections between the discrete experiences of its members. For Dos Passos the primary issue was not the photograph's ability or inability to record unfathomed depths or hidden relationships, but rather the opposite. To him, relationships did not seem to exist. In this sense the anomie of the artist, separated as he was from the life he chose to observe, became the emblem of an entire culture's alienation from itself. No more connected to the world of his own observation than Nathaniel Hawthorne had been, Dos Passos nonetheless gave to his own experience a collective and emblematic status. Where Hawthorne had pictured the artist as a voyeur posed outside of a lively and coveted world he could approach only through vision, Dos Passos posed the artist as an outsider in an entire culture of outsiders. A farmhand who comes to town at the beginning of *Manhattan Transfer* asks "How do I get to Broadway? . . . I want to get to the center of things." But it was Dos Passos's opinion that Manhattan had no center; it was only a series of peripheries, an aggregate of unconnected, "pigeonhole" lives.

Dos Passos's most remarkable and remarked-upon narrative strategy in *Manhattan Transfer* was a dispersal of focus, a refusal to privilege any one character or to accord especial weight to an individual sensibility. What Philip Rahv has said about the naturalistic novel as a genre is certainly true of this example of Dos Passos's urban fiction—that the main character is the environment, the city itself whose moods are revealed, whose changing aspects are made known through the destinies of many insignificant people. It was a technique adopted to reflect the fragmentation of urban life: Bud Korpennig is a vagrant from the country who never makes any place for himself and whose past—he has killed his father—coupled with the city's indifference leads him to suicide. Ruth Prynne and Cassie Wilkins are nondescript girls who aspire to, but do not succeed in, stage careers. Anna Cohen is a factory worker who eventually is disfigured by a horrible fire. Francie and Dutch are nice kids and drifters who pull off an armed robbery; Rosie and Jake are the con-artists behind the Prudence Promotion Company. Joe Harland has lost his place on the stock market and is, like Dreiser's Hurstwood, a Bowery bum with memories of a grandiose past. George Baldwin, a lawyer, makes his money by suing the utility companies; he sleeps with Nevada Jones, a prostitute, who marries Congo, a booze runner. James Merivale tends his own wealth and spends his time at the Metropolitan Club; his sister, Maisie, marries Jack Cunningham despite knowing that he is already married to Jessie Lincoln. I have drawn out this list in order to make a rhetorical point, for its length and diversity constitute its meaning. To the extent that the novel leans toward any of its characters, it favors Ellen Thatcher, a spoiled, attractive, and increasingly callow woman whose stage career and serial marriages preclude any real attachments, and Jimmy Herf, who marries her briefly, dabbles at journalism, has anxious dreams, and, like Huck Finn, ends the novel by heading out "for the territories."

Manhattan Transfer is obviously a young man's novel of disillusionment; it is written in recognition and defiance of the distortions of urban, industrial culture. Dos Passos's New York is a cacophony of shrill, ruined lives; the prose is nervy, grating, in imitation of the radios, taxis, and fire engines that bleat in the novel's background.

There is no quiet, no stopping, no self-reflection in the book, and Dos Passos equates the scene with Ninevah before its final destruction: "Thus saith the Lord, he that remaineth in this city shall die by the sword, by the famine and by the pestilence: but he that goeth forth . . . shall live."[2]

Any moralism in the text remains implicit, inferable from the waste and disasters reported, but not from the narrative voice itself: Dos Passos remains, like his protagonist, Herf, the reporter of lives, not their judge. This is an appropriate stance, for it is the culture that is indicted, not any one of those struggling within it; the very multiplicity of perspectives, borrowed though it may be from Joyce, is called into use to add to a collective picture. What is of interest in the novel, and what Dos Passos recognized as a major creative dilemma, is the relations of parts within the larger structure: "Was it sufficient," he asked, "for the artist to mimic dislocation and brittleness, to use imitative form as a sole artistic principle, and to remain, by virtue of that form, the artist as a reporter rather than a synthesizer or prophet?" The young journalist of the novel continually reenacts Dos Passos's own creative dilemma as he wanders around the city on assignment, seeing, but never entering into, the human dramas that end up in newspaper columns. When he happens to be at Congo's during a bootleg liquor bust, he stays inside:

> I ought to get out and see what's going on. He groped for the front door. It was locked. He walked over to the piano and put another nickle in. Then he lit a fresh cigarette and started walking up and down again. Always the way . . . a parasite on the drama of life, reporter looks at everything through a peephole. Never mixes in.

Later he lacerates himself again when a friend asks him if he's going to shovel snow for the city: "I don't know why it wouldn't be better than spending all your life rooting into other people's affairs until you're nothing but a goddamn dictograph."[3]

These denunciations were, quite simply, the criticism Dos Passos had leveled at himself during his early manhood. To his companions, he alternately raved about New York, its "wonderful atmosphere of gaiety and a sort of paganism," and admitted his own problematic relation to that "paganism."[4] He would sometimes find himself in "the horrid state (not unknown to me) in which—from pure watching of other people, from trying to pry into their lives as they pass me on the street—I have reached a point where I feel more like a disembodied spirit than a warm-fleshed human."[5]

He was, quite clearly, reexperiencing Hawthorne's artistic

dilemma, finding himself—either by natural inclination or because of the dictates of his art or both—fixed into a seeing/being dyad, standing outside in order to gather together the raw material of eventual texts. Edmund Wilson remembered him in a similar fashion in his novel, *I Thought of Daisy,* where the fictionalized Dos Passos "'was really on close terms with no one.' Wilson thought that Dos Passos was the kind of man to sample a conversation to catch 'the social flavor of a household or group'; then he would catalog the sample, placing his last specimen where he thought it belonged in the economic system."[6]

What is unique about Dos Passos's experience of alienation is the placement of this experience in his early novel, for he attributed these feelings of anomie to everyone in New York and thought them to be the result of living in an economic and political system that was heavily buttressed by images of unattainable glamor. Jimmy Herf is not, in Dos Passos's novel, a privileged and uniquely suffering character. He is one of a group of young people who all experience variations of voyeuristic emptiness and who find that the "center," which they hope to approach through seeing images of it, does not exist. It is as if Dos Passos sets up a seeing/being dyad only to demolish the second term so that empty hopes and visions are all that remain of life. Each of these characters, in ways appropriate to their origins, approaches "the good life" that post-war Manhattan seems to promise them in ads and on showbills—and each is rebuffed. Jimmy grasps this vaguely when he says anxiously, "Don't talk. . . . What you talk about you never do."[7] In some still inchoate way, he has understood there to be an inverse relationship between action and the representation of action in language or in image, as if the existence of the word, the image, could, in itself, sap vitality, rob the real world of its own inviolate energy. Later in his life, Dos Passos would make this connection more explicitly, when, in *Century's Ebb,* he asked, "Can it be that the Arabs are right, and the dour pueblo-dwellers in our own Southwest, when they say that the camera takes away something that can never be recovered, skins some private value off the soul?"[8]

At this early stage, however, there is only a glimpse of this thought, a pale anticipation of the later vehemence with which he would denounce America as an "image nation," a country manipulated and subdued by photographs and by its own pictures of desire. Here, in *Manhattan Transfer,* Dos Passos is still playing with the idea of psychological vacancy, understanding it to be symptomatic of urban life but not yet locating the origins in a social structure or political practice. The character Ellen Thatcher, whose marriage and affairs carry her farther and farther from joy, is his foremost example of glamorous

emptiness; and as her career progresses, she becomes increasingly fran-
tic, and increasingly numb. Dos Passos expresses her distress as a kind
of absence as if she were only an image of her former self: "Through
dinner she felt a gradual icy coldness stealing through her like novo-
caine. She had made up her mind. It seemed as if she had set the photograph of herself in her own place, forever frozen into a single gesture."[9] What turns this one experience, expressed through an incidental
photographic metaphor, into an incident worthy of deeper consideration is its typicality. To Dos Passos, the photograph as a "skin," a "surface," a reference to something absent, a fragment, expressed something
widespread and profound: to be or to feel like a photograph of oneself
was to be caught in, or have accepted, someone else's image of what
one should be. It was the ultimate expression of alienation, the substitution of externally imposed norms for internally motivated action, the
eclipse of innate character. Everyone in Dos Passos's New York has a
tendency to turn him- or herself into a photograph, metaphorically as
in Ellen Thatcher's example; literally as in James Merivale's fantasy of
a report of his own fame: "Ten Million Dollar Success . . . (flashlight
photograph) . . . (flashlight photograph)"; or to live in such a way that
newspaper photographs have as much reality as flesh and blood:

> [Mrs. Cunningham] was looking at a photograph in a rotogravure
> section labeled Mr. and Mrs. Jack Cunningham Hop Off for the First
> Lap of Their Honeymoon on His Sensational Seaplane Albatross
> VII. . . . Mrs. Cunningham heaved a deep sigh and settled herself
> among the pillows. Outside churchbells were ringing. "Oh Jack you
> darling I love you just the same," she said to the picture. Then she
> kissed it. . . . Then she stared at the face of the second Mrs. Cun-
> ningham. "Oh you," she said and poked her finger through it.[10]

In these examples we can see that Dos Passos has collapsed the photographic process, making everyone the photographer of himself, as if all
of life were to be experienced through the viewfinder of a camera.
There is no subject. To observe oneself acting, to become self-conscious
in this way, is to violate the roots of authentic action. And if the self
does not exist, then the other is equally invisible: "His wooden face of
a marionette waggled senselessly in front of her." To Dos Passos, this
formulation of the dynamic of self and other, the substitution of the
outside view for inner direction, was the sign of civilization's bank-
ruptcy. The burning buildings, the sirens, the sensational distresses of
the city coalesce, building to a cumulative picture of horrendous emptiness. Worship of the isolated image of the self is Dos Passos's version
of idolatry; it is Ninevah's false god that Jeremiah prophesies will be

destroyed in wrath. In 1927, Dos Passos could name the cultural destitution he saw without being able to extricate himself from it. The artist
was, as he knew, caught in the dangers of observation as much as anyone he wrote about; the artist was, in fact, the prototype of these alienated lives. Don Passos's advantage was another kind of self-reflection
which let him know of his own malaise; but in *Manhattan Transfer* he
was only able to find narrative form for the negative thing, for fragmentation, and the lack of connection.

II

In his next fiction, *USA,* Dos Passos was to find formal expression for
a more complex rendering of American history and to understand the
artist as someone who could intervene in that history as well as watch
it. He found the seeds of this new understanding in the Soviet Union.
Through meeting with film directors, seeing new work, reading their
manifestos and theories, he came to see the nature of his own task more
clearly. Their raw material, like his own, was the cultural fragment,
their aim, as his own came to be, was to weld "into a sentient whole
the rigid honeycomb of our pigeonholed lives." In Russia, he said, they
"adore the film as if it were a god."[11]
    He went there in 1928, and though he had already been at work
on *The 42nd Parallel,* the book that would be the first volume of the
USA trilogy, there is little doubt that his contact with Soviet film directors affected the narrative stance, and, indeed, the entire structure of
his subsequent work. In essence, they gave him a way to envision the
relation of part to whole, to see the historical moment or the isolated
life as part of a coherent process. Dos Passos eventually took an artistic
form that had been created to illustrate the way in which contemporary
Russian history embodied a Marxian dialectic and applied it to the
description of capitalist society: where the goal of the socialist film-
makers had been to give concrete expression to the synthesis of opposing factors that the October Revolution had already effected, Dos Passos's aim was to show the post-industrial world of America posed at a
moment before synthesis, shattered by oppositions or "contradictions"
that foreshadowed and demanded similar social action. He ended the
trilogy with the admission, "[W]e are two nations." But by 1936 he
could see these divisions in the context of a potentially greater coherence. If *Manhattan Transfer* had pictured a tormented and fractured
world whose end was prefigured in Biblical history by the destruction
of Ninevah, *USA* showed those same fragments to be subject to the

analysis of more immediate human history—an analysis created by men he could talk to, theories he could espouse, examples he could and did emulate.

In his memoirs and letters, Dos Passos remembered the art of the Soviet Union primarily in terms of its theater. He mentioned V. E. Meyerhold's *Roar China* as a magnificently staged production, and he was moved even more by the workers' extemporaneous, autobiographical productions which they called "Living Newspapers." In light of his own involvement with the New Playwrights and the radical theater scene in New York City, this is not surprising: he was watching the successful execution of an enterprise that had failed on American soil. But to e. e. cummings he later wrote that the "most interesting and lively people" he had met were the film directors, and he singled out Sergei Eisenstein as someone with "one of the most brilliantly synthesizing minds"[12] he had ever encountered. Eisenstein has, consequently, been generally regarded as a major influence on his writing. A synthesizer of fiction and history, proponent of montage and contrast as organizing principles of the film, theorist of art's role in the broader arena of political action, Eisenstein was undoubtedly important to Dos Passos—an opinion that is corroborated by Dos Passos's retrospective accounts of his own creative development. But I suspect that his written admissions cloak the more direct or disturbing influence, that he refrained from speaking about the most important and controversial origin of his narrative technique. For Dziga Vertov, the film director whose own manifestos were called "The Cinema Eye," must certainly have been the source of Dos Passos's "The Camera Eye" in *USA;* and Vertov's film *The Man with the Movie Camera* must have been the one single work of art whose structure suggested to Dos Passos the purpose of the artist in history, the function of self-dramatization in art, and the relation of subjectivity and objectivity. These issues, which provide the center of Vertov's profound meditation on the role of art in Soviet culture, were the very issues that Dos Passos had found most problematic in his early fiction and whose relation had remained unresolved in *Manhattan Transfer.* He had motive to see Vertov's work, the opportunity to see it, and a further motive for disguising what might later detract from people's admiration for his own originality. In obvious ways, "The Cinema Eye" and "The Camera Eye" were too closely related for Dos Passos's aesthetic comfort.

What films could Dos Passos have viewed during his travels? We know, because he mentions it, that he saw *October,* Eisenstein's rendition of the great revolution. It was released in May 1928, as was Vertov's treatment of the same historical events in a film called *The Eleventh*

Year. When Dos Passos returned to Moscow in November of the same year, there was another series of releases: *The White Eagle* (directed by Protazanov, released 9 October), *Eliso* (directed by Shengalaya, released 23 October), *The Heir to Jenghis-Khan* (directed by Pudovkin, released 10 November), *The Russia of Nikolai II and Leo Tolstoy* (edited by Esther Shub, released 10 November), *Katorga* (directed by Raizman, released 27 November), and *The Man with the Movie Camera* (directed by Vertov with previews which began on or before 7 September).[13] What is of equal speculative relevance to Dos Passos's Russian visit are the "manifestos" that Vertov circulated at the time in the early *Kino Pravdas*. Though many film makers might have agreed with Vertov's "death sentence" to all films made before 1919, though they might have concurred with his belief that western models could no longer provide direction for the social realities of the new Soviet state, few of them actually subscribed to his flamboyant experimental pronouncements about art and film technique. In fact, by 1928, Vertov had incurred considerable official disapproval and, having lost his position at Sovkino in Moscow on 4 January 1927, he was filming in the provinces at VUFKU (the All-Ukrainian Photo-Cinema Administration).[14]

Regardless of his controversial status, or perhaps it would be better to say because of it, Vertov was, in 1928, a highly visible Soviet artist and had been since 1918 when he first had been given work making newsreels by the Moscow Cinema Committee. The circumstances of his early work life explain, to a large extent, the aesthetic commitments of his later career, for the revolution both established the goals of his cinematography and imposed limitations upon it. Vertov's genius was in creating a theoretical bridge between the ideological imperatives of Lenin's regime and the material restrictions placed upon his work: without adequate film he was asked to inform a huge, disparate population of the nature of wide-ranging social change, the contribution of specific events to that change, and the significance of individual lives within the shifting structure of the whole. He quickly defined his own role on the documentary team as that of editor: working with film clips from many cameramen, dealing with seemingly unrelated events from different parts of the country, he synthesized, forming fragments of raw material into a comprehensive and coherent picture of national purpose. Historians of the film have attributed many of Vertov's techniques to the influence of the Cubo-Futurists, that is, to the pre-revolutionary avant-garde movement in Petrograd, which emphasized the speed, power, and mobility of new technologies,[15] but I am convinced it was historic necessity that influenced him most: Vertov's lack of film and his subsequent need to use a small unit of production gave to his

work a form that paralleled the imperatives of the Revolution itself.
The compilation of detail, the imposition of coherent purpose on actual
events that may have seemed chaotic or senseless in themselves, were
the tasks of Lenin in building a radically new social order, and of Vertov in making a film to represent that order. The central structural
problem for both was the relation of parts to a whole built cumulatively. For Vertov, metonymy, the part standing for or representing the
whole, was never a serious, aesthetic possibility: to think of a representative individual or typical event was to misconstrue social truth which
remained, always, in the totality. Montage was, then, initially a necessity imposed by lack of materials; it became a mimetic form, and to
later critics it seemed an avant-garde posture. Eisenstein, who also
thought deeply about the appropriate form for socialist film, came to
think of Vertov as concerned with tricks of the camera, with speed or
change for its own sake. What interested Eisenstein most profoundly
was the film's capacity to establish and then resolve opposites, as if it
were a pictorial equivalent of Marxist historiography, as if it were
emblematically synthesizing thesis and antithesis and showing the
Hegelian structure underlying reality. With his alternative theory for
the representation of history in film, Eisenstein did not condone the
more amorphous structure of Vertov's work. But it must be said in
Vertov's behalf that although he, too, worked mimetically, he did not
represent a dialectic; he represented instead the lack of clarity which
faced him in 1918 as he rode through the actual battles of the Civil
War and worked in studios improvised in moving trains. To Vertov,
socialist reality did not seem an orderly dialectical progression of thesis
and antithesis, but something more accurately recorded as a simple collection or series.

These are, of course, speculations. What is clear is that Vertov
remained committed to those techniques that characterized his early
work as a documentarian of the Revolution: years later we find him
reiterating the general theory of montage as a technique for dealing
with fragments and theme:

> But it is not enough to show bits of truth on the screen, separate
> frames of truth. These frames must be thematically organized so that
> the whole is also truth. This is an even more difficult task: there is
> little theoretical study of this problem. Hundreds, thousands of exper-
> iments must be conducted, in order to master this new field of cine-
> matographic work.[16]

He then refined its nature:

Montage takes place from the beginning to the end of production. . . . *First Period:* The "Montage Evaluation" of all the documents that are directly or indirectly related to the chosen theme (manuscripts, various objects, film clippings, photographs, newspaper clippings, books, etc.). As a result of this montage, which consists in picking and grouping the most precious documents or those simply useful, *the plan indicated by the theme* becomes crystallized, appears more evident, more distinct, more defined. *Second Period:* "Montage Synthesis" *of the human eye* concerning the selected theme (montage of personal observation or of reports by the information-gatherers and scouts of the film). *Plan of shots*, as a result of selection and classification of the observations of the "human eye". . . . *Third Period:* "General Montage," synthesis of the observations noted on the film under the direction of the "machine-eye. . . . " Unification of homogeneous pieces; constantly, one displaces the pieces, the frames, until all shall have entered a ·rhythm, where all the ties dictated by the meaning shall be those which coincide with the visual ties. . . . This formula, this equation, which is the result of the general montage of the cine-documents recorded on the film, is 100 per cent the cine-thing: I see, I cine-see.[17]

In the same way that the individual frame contributed to the final theme of the film, the individual life was regarded by Vertov as important only insofar as it participated in the building of the socialist state. Even Vertov's treatment of Lenin in *Three Songs of Lenin* adhered to this principle and showed the Soviet leader not as a great personality but as someone who contributed to institutional life that would benefit the collective. Vertov regarded "personality" or emphasis on individual, affective life as a bourgeois indulgence. Insofar as he dealt with emotional life at all, he treated it as a matter of fact: the human being harbored no hidden complexities, no contradictions, no concealed longings or desires. Character was available for everyone's perusal, and since it was revealed on the face, it was a subject easily accessible to the camera. The experiment Vertov made early in his cinematic career allows us to see and understand the kind of information he thought his film able to register. Jumping from a second story balcony, he had the action recorded on film in slow motion:

The results. From the point of view of the ordinary eye it goes like this: the man walked to the edge of the balcony, bowed, smiled, jumped, landed on his feet and that is all. What was it in slow motion? A man walks to the edge of the balcony, vacillating. To jump or not to jump? Then it is as if his thoughts say that everything points to the need to jump. I am entirely uncomfortable. Everyone is look-

154 IN VISIBLE LIGHT

ing at me. Again doubt. Will I break a leg? I will. No, I won't. I must
jump, I cannot just stand here. An indecisive countenance is replaced
by a look of firm decision. The man slowly goes off the balcony. He
is already situated in mid-air. Again, fear on his face. On the man's
face are clearly seen his thoughts.[18]

What is interesting about Vertov's experiment, aside from his reckless-
ness, is his absorption in the action itself: turning the camera on himself, in front of instead of behind the lens, he experienced, or at least
claimed to experience, no emotions extraneous to the fear and hesitation appropriate to the jump. Yet it was done in full consciousness of
the camera. Was it or was it not a performance? Vertov's critics have
pointed out that any behavior elicited by the camera should be cate-
gorized as a performance; but the more important point is really that
Vertov thought it possible to be unaffected by the camera's perusal, to
avoid the deflections of "normal" conduct that we usually associate
with being watched. And however unusual we may find this, it was not
an unreflective attitude. In fact, Vertov's meditation on the camera in
the midst of life led him to believe that it could actually assist in class
analysis and in the evaluation of social commitment. If one were totally
involved in one's work, Vertov postulated, one would be unaware or
undisturbed by a photographer's presence; one would simply continue
working. It was only those who persisted in holding bourgeois attitudes
about appearance and who had not given themselves entirely to their
labors who would experience discomfort or embarrassment in front of
the camera. Hence the "unplayed" documentary served two functions,
one a subsidiary of the other: it rendered the entire truth of a social
reality, and within that larger truth, it allowed one to distinguish
between those dedicated to socialism and those who still held private,
decadent opinions. In *The Man with the Movie Camera,* the woman who
mimics the cranking wrist action of the cameraman is clearly a throw-
back of western culture; she is someone extraneous to socialist purpose,
a useless person.

Part of Vertov's insistence on understanding the meaning of self-
consciousness before the lens grew out of a larger reflection on the role
of cinema within the culture at large, a reflection, or set of reflections,
that was addressed in *The Man with the Movie Camera,* his most famous
and controversial film. It is clipped, edited ferociously, and was widely
condemned by Vertov's contemporaries as a decline into pure formalism. In that the film is both cinema and meta-cinema, at once a comment on Soviet life and on the recording of it, it is formal; in that the
frenetic pace of the film calls attention to itself, it is formal; but in that

The Man with the Movie Camera remains a kind of exuberant visual polemic about the photographer as a productive laborer whose work is analogous to the industry he films, it is fully consonant with Vertov's earlier views about the relationship between politics and aesthetics. The function of montage still is clearly to confront images with each other, to juxtapose, and by contiguity, to show analogy: school children, housewives, factory workers, the cameraman, the film editor, and the film's audience are all similarly engaged in a social project where creativity and productivity are comfortable with each other—they all produce meaning collectively. Thus, shots of the cameraman are repeatedly intercut with shots of foundry workers, travelers, athletes, motorcycle racers; shots of the film editor handling rolls of film are intercut with machinery with swirling spindles. Shots of citizens in daily life are interspersed with shots of others in the theater, watching their peers. In whatever ways Soviet life of the 1920s went on, it went on with the photographer in the middle of it—filming trolleys, policemen, traffic, machinery, or whatever. This was clearly Vertov's way of trying to show the integration of innovation and ordinary life, his plea for the audience to accept as common a way of looking at the world and themselves that was, as others rightly observed, decidedly uncommon. In his own mind, there was a clear distinction between innovation and idiosyncracy. One was a tool of liberation, the other a sign of decadence. One was used in the service of political orthodoxy, the other was a bourgeois anachronism. He wanted others to see that the artist, the innovator, belonged in the same sequence of frames as everyone else. That his critics failed to understand the difference between creativity that originated in social vision and creativity that was a manifestation of the artist's privileging of his own personality must have disappointed him, for Vertov never suggested that socialist art should be socialist realism in the sense that realism is commonly understood to imply a commitment to undistorted representation. Nor did he claim that it should be the direct vehicle of ideological teaching. In fact, the excitement of art, for him, lay in the combination of the unplayed scene, the honest and unrehearsed perspective of the camera apparatus itself. To Vertov, the camera remained a machine; it was decidedly not a mere extension of the photographer's eye, and its virtues were those of the machine, not of the naked and unassisted observer. For this reason, the camera, alone on its tripod, was as much a character in *The Man with the Movie Camera* as the director, editor, or photographer: it was a visual embodiment of Vertov's pronouncements about "unserfag[ing]" the camera, freeing it to see in ways that the human eye could not:

The utilization of the camera, as a cinema eye—more perfect than a human eye for purposes of research into the chaos of visual phenomena filling the universe.

The eye lives and moves in time and space, perceiving and recording impressions in a way quite different from the human eye. It is not necessary for it to have a particular stance or to be limited in the number of moments to be observed per second. . . .

To this day, we raped the movie camera and forced it to copy the work of the eye. And the better the copy, the better the shot was considered. As of today, we will unshackle the camera and will make it work in the opposite direction, further from copying.[19]

In these attitudes he anticipated Lázló Maholy-Nagy's interest in the camera and in photographic vision. Like Vertov, he thought the camera's great virtue to be its intensification and alteration of vision—its ability to change the speed, size, depth, and abstractness of things in ways that eluded the eye, permutations he eventually codified in "The Eight Varieties of Photographic Vision." Like Vertov, he also saw another excellence of the camera in the idea of the series, an idea he called "the logical culmination of photography."

The series is no longer a "picture," and none of the canons of pictorial aesthetics can be applied to it. Here the separate picture loses its identity as such and becomes a detail of assembly, an essential structural element of the whole which is the thing itself. In this concatenation of its separate but inseparable parts a photographic series inspired by a definite purpose can become at once the most potent weapon and the tenderest lyric.

He added, "The illiterate of the future will be ignorant of the use of the camera and pen alike."[20]

It was an indictment that never would have touched Dos Passos, for he worked on notes to *USA* while in Russia, and he returned to the United States with plans for his own version of "The Cinema Eye" intact. In the modifications of narrative technique, in the differences that characterize the writing styles of *Manhattan Transfer* and *USA,* we can read Dos Passos's susceptibility to the influence of Soviet film. For Vertov, Eisenstein, and Pudovkin were engaged in a debate which addressed many of the problems that had remained unresolved in *Manhattan Transfer:* in the early novel, Dos Passos had recorded the experience of an aimless and alienated collection of people whose lives never coalesced. He cursed them, as Ninevah had been cursed, but since he considered the artist as alienated as those he observed, he saw no function for himself other than that of lame prophet. By the time he wrote

USA, Dos Passos had understood his own artistic role in a much fuller sense and had begun to use Vertov's interval theory to make fragmentation a virtue—both as a continued mimetic device and as a principle that made disparity into an active contribution to a larger coherence. Dos Passos went far enough to see coherence as a characteristic of fiction but not as part of the historical world to which the text alluded. For him, it was precisely the lack of national purpose in America that gave him his most urgent theme, his most serious task. Though he clearly referred to himself as a socialist writer in "The Writer as Technician," the piece he wrote for the First Writers' Congress in 1935, he also knew that the arts were not reinforced by national policy in America as they had been in Russia.

It is not certain that Dos Passos saw the first book in the trilogy, *The 42nd Parallel,* as the beginning of a three-part novel. He did, we know, have something of epic proportions in mind, a saga of post-World-War America as broad and all-encompassing as Vertov's *One Sixth of the World* (1926). Where Vertov tried to define the Soviet Union by addressing its many ethnic groups as if they were a catalogue—"You in the small villages," "You in the tundra," "You on the ocean," "You Uzbeks," "You Kalmiks"[21]—Dos Passos defined America in terms of more amorphous but nonetheless representative types. America's diversity was due not so much from ethnic groups—though initially that may have been so—as from poorly defined and even more poorly acknowledged classes. In Dos Passos's world, money, prestige, and their consequent opportunities are the characteristics that distinguish one person from another. His version of America is composed of secretaries (Janey) and enlisted men (Joe Williams), of midwestern bitches (Eleanor Stoddard), of opportunistic young studs (Richard Savage), of random, lucky mechanics who jockey their tinkering ability into fat Wall Street careers (Charlie Anderson), and of glib talkers who wheedle products into households and package government policies for international consumption (J. Ward Moorehouse). One of them sleeps her way into Hollywood notoriety (Margo Dowling), another drowns her pain first in a nice little marriage with babies, then in lovers, then in suicide (Eveline Hutchins). A few of them are "working class stiffs" involved in the American Labor Movement; others are middle-class idealists who try to understand the social unrest that seems to impinge obliquely on everything. Dos Passos's themes are imperialism, socialism, war, class struggle, and the subjective manifestations of a social order in radical disarray: "Whoring, violent drinking, and constant aimless motion."[22]

Like *Manhattan Transfer,* this new work was constructed from tiny narrative blocks: he told no life story in its entirety but instead broke each into segments which intercut each other. These, in turn, were intercut by other kinds of narrative the variety of which has caused some critics to call the trilogy a "contraption," and to see Dos Passos, like Vertov, as a formalist concerned with aesthetic pyrotechnics. Alfred Kazin called the book "an aesthetic proposition about style in relation to the modern world,"[23] but it is certainly not a proposition about style alone. Like his predecessor in film, Dos Passos used narrative structure to embody an understanding of history, and like Vertov, he was preoccupied with the relationship of the common person, the artist, and the great sweep of historical events. In the trilogy, the lives of common people, the Janeys and Charlies of Dos Passos's first narrative mode, are juxtaposed to the lives of those of great achievement or notoriety: Thomas Edison, Thorstein Veblen, Frank Lloyd Wright, and Isadora Duncan are examples. All of these lives are contextualized by "Newsreels" made from the contemporary newspaper clippings, songs, clichés, headlines, and advertisements. These three modes, the life stories of common people, of famous people, and "Newsreels," run alongside a fourth, "The Camera Eye," which breaks the narrator's own life into fragments and recounts personal history as an element of commensurate weight. The whole narrative has an audacious, rushed, and despairing quality, as if speed itself could lend relief to destitution.

In one sense, USA reiterates and expands Manhattan Transfer; in another sense, it violates or challenges the entire narrative logic of the earlier work: in one, the bricolage is local, in the other, epic; in one a lack of connection is the point, in the other, the fragment is the prelude to synthesis, to construction of the historical world, and to understanding the ways in which the broad movements of history impinge on and affect individual consciousness. Alfred Kazin has said that the book is "an enormous chronicle of disillusionment,"[24] because history overpowers the individual. But it is equally possible to argue that Dos Passos was relieved to discover a mode of representing history, relieved to understand that isolation could be contained within a context. In Manhattan Transfer, the problem he addressed had been fragmentation or the "pigeon-holed" lives of those who could understand none of the historical causes of their estrangement. The function of the novel was to present the problem itself; it was a novel of consequences. The Soviet example gave Dos Passos a way to use the fragment, the small structural unit that was already his preferred narrative mode, and to activate it for the audience. From Vertov he took the idea of the interval, the thought that the space between fragments could invite participation, that the film-maker/writer/technician's job was to edit, to provide the juxtaposition of information that, when assembled in the viewing/reading, would lead to a recognition of the importance of each unit within the whole. The function of art, in this formulation, was to present a picture of wholeness, to show the effects of history itself. Even though the individual depicted within the whole might not understand the parameters of his own existence, the audience, through the artist/editor's guidance, would.

This commitment to the structural whole underlies and explains Dos Passos's treatment of individual lives within the text. Critics have noted (and decried) Dos Passos's "objectivity," his apparent lack of sympathy for the lives of those he reveals, without, I think, understanding the seriousness of his struggle with the meaning of such objectivity. His distance from his subject, his seeming voyeurism may have originated in psychological discomfort, but it did not remain a simple personal inclination. In fact, Dos Passos, more than any of his predecessors, came to understand the political ramifications of objectivity and to use it in the service of a comprehensive vision. Henry James had touched on the social implications of voyeurism, of the artist's estrangement from the field of observation, by making the artist/observer of *The Princess Casamassima* an elitist with an interest in the lower classes of London. Dreiser had a similar sense of the creation of art as a reiteration of other patterns of class interaction. Dos Passos made a clear distinction between two kinds of objectivity, and he reenacted both in his own career. The artist who merely looked on, who took an aesthetic interest in the lives around him but who remained the voyeur, was a bourgeois artist. Jimmy Herf in *Manhattan Transfer* was this—the estranged aesthete whose only solution, exile, was no solution at all. The artist of *USA* is someone with an entirely different understanding of the nature of objectivity, an understanding that is a consequence of seeing himself as a participant in history, a worker who has overcome his own isolating subjectivity. The artist in *USA* puts things together; he creates meaning by assembling. He is an overseer of history and his text assumes a social dimension by providing the means of audience/reader participation in that overview. One could say that Dos Passos's *USA* becomes the metaphorical equivalent of the movie theater, the viewers of which were dramatized in Vertov's film, *The Man with the Movie Camera*. And like Vertov in that film, Dos Passos chose to dramatize himself, the artist, within his own work: this is the function of the "The Camera Eye" within the trilogy: its placement, as a series of fragments within other narrative fragments, is Dos Passos's equivalent of Vertov's dramatization of the cameraman in the film: just as the photographer is a worker among workers and photography is a labor commensurate with other kinds of industry, so the writer/perceiver has a status equal to the lives of those narrated: the intercutting of parallel structure tells us this. But "The Camera Eye" sections also chart the growth of the artist away from isolation within the confines of the self toward the ability to take an outside view that was not based on personal estrangement. The objectivity that interested Dos Passos at this stage in his career was a tool of a larger commitment and purpose. Later when he worked with Joris Ivens in filming the Spanish Civil War, he came to believe this posture of facticity, of photographic accuracy, to be dishonest, an ideological screen; but in *USA* he saw it as a way out of a previous dilemma, as a way to "declass" himself and to define art's function and his own social usefulness.

III

The early drafts of "The Camera Eye" sections of *USA* are undated; but before Dos Passos began to write any of them, he tried working out the general parameters of the sections:

> The upside down image in the retina, piece by piece immediately out of color shape remembered bright and dark rebuilds the city [Sunday] sunlight on the downtown streets [stained even] pavements—Truck cluttered streets—Terrible dead city on the make—[25]

Another version went:

> Camera Eye—the careful clipping out of paper figures—the old photographs in a trunk the pathetic enthusiasm—how could you wrap the paper figures to simulate growth—twist cut out a pack and [tickle] whittle it up to eighty—shove him through the terrible various velocities of time. Imagine paper boats that will indicate the swirls and eddies of the stream.[26]

Whether Dos Passos actually read the various manifestos, creative plans, and testimonials that Vertov published in *Kino Pravda* during the 1920s is something that we can never know with absolute certainty. But the circumstantial evidence is strong: he had the opportunity to read them, and his own tentative schemes for "The Camera Eye" echo the language and concerns of Vertov almost verbatim. The cinematographer spoke of the intervals, the relation between the pieces of the montage as "a great complexity, formed by the sum total of the various relations of which the chief ones are: 1) relations of planes (small and large); 2) relations of foreshortenings; 3) relations of movements within the frame of each piece; 4) relations of lights and shades; 5) relations of speeds of recording."[27] When we look at Dos Passos's first efforts to formulate "The Camera Eye" sections of USA, we see that he worked with all of Vertov's categories: his camera eye will operate in terms of "bright and dark" (Vertov: "relations of lights and shades"), in terms of "velocities of time" (Vertov: "relations of speeds"), of "growth," or "swirls and eddies" (Vertov: "relations of movement"), and "piece by piece immediately out of color shape" (Vertov: "relations of planes"). What is most important, though, is Dos Passos's general understanding of the principles that led Vertov to formulate his interval theory in this way—his conviction that the camera should not be used to duplicate human vision but to create that which the unassisted eye could never see. The excitement about the camera for Vertov, for Maholy-Nagy, and eventually for Dos Passos in his own medium, was that it was a mechanism that could evade the requirements of normal sight, for in their view the "simple eye" represented the persepctive of all those individuals who could not maneuver through their own limited subjectivity, who could not see or experience the broader historical world of which they were a part. To these artists, the single point of view was politically inadequate; they attributed their own greater clarity, in part, to their familiarity with the camera itself, to the machine: "[F]reed from the frame of time and space, I coordinate any and all points of the universe, wherever I may plot them. My road is toward the creation of a fresh perception of the world. Thus I decipher in a new way the world unknown to you."[28]

Dos Passos eventually modified his response to Vertov; his final sections of "The Camera Eye" bear only tentative relation to these early schemes, and, in fact, what he first talked about doing in "The Camera Eye" segments more adequately describes the plan of the whole trilogy than does the treatment of his own life or the artist's perspective. It is as if Dos Passos sifted through these influences, backed away from them, and then produced a narrative structure that stood only in the most broad relation to the ideas of his predecessor. He did not simply translate film technique into prose. When he did sort out what to do with "The Camera Eye" in USA, he understood that it should have a more specific purpose; and though we can say that Dos Passos used a time and space montage—collapsing time and making creative juxtapositions—he was primarily interested in the meta-fiction that the sections enacted—the opportunity to chart the genesis of the speaking voice of the text within the text itself.

It is here that we come to understand the ways in which Dos Passos "edited" himself, seeing his own growth from childhood through the disturbances of adolescence and the traumas of World War I, the Sacco and Vanzetti trial, and the labor violence and destitution of the mining strikes in Harlan County, as a growth from solitude and protection *from* history to collective identity and another kind of alienation that should be understood as the result *of* history, a certain perversion of ideals of the original American charter, as Dos Passos understood that charter. Where he himself began life as a foreigner having to fight for his status as an American ("Hey Frenchie yelled Taylor in the door you've got to fight the kid doan wanna fight him"),[29] he moved through literal enfranchisement—Choate, Harvard, the American Ambulance Corps, and the military—only to see himself as a foreigner again, estranged by the very qualities of American life that he had chronicled in the contiguous sections of the trilogy—the Newsreels. America's "songs and slogans, political aspirations and prejudices, ideals, hopes, delusions, frauds, crackpot notions out of the daily newspapers"[30] were not his own. What is important to remember is that the estrangement of *USA* is not that of the aesthete; it is not a reiteration of the isolation and disillusion of Jimmy Herf in Dos Passos's previous work. The aliena- tion here is expressed in a collective voice: "they have made *us* [my italics] foreigners in the land where we were born they are the con- quering army that has filtered into the country unnoticed they have taken the hilltops by stealth."[31] While the young man whose life is charted in "The Camera Eye" sections knows solitude, knows lonely streets, knows how to dabble at politics, watching radicals (Max East- man as he speaks at the Garden or Emma Goldman as she eats hot dogs and sauerkraut in restaurants), he does not remain the voyeur, but comes, instead, to understand his own betrayal as part of a collective experience. Like Herf, whose alienation came to be emblematic of the alienation of an entire generation of urban youth, the narrator of "The Camera Eye" also has the emblematic identity of a group, but this self- definition does not rest with identifying his "pigeonholed" life among other unfeeling solitaries. He speaks collectively; he addresses America as a corporate structure; he has forsaken an earlier subjectivity for a larger view of the self in history: "Make them feel who are your oppressors America rebuild the ruined words ... how can you know who are your betrayers America or that this fish peddler you have in Charlestown Jail is one of your founders of Massachusetts."[32]

The distinctions between Dos Passos's and Vertov's senses of the artist's position in history are numerous and obvious. Vertov considered the photographer to be a worker among other workers, all of them engaged in the construction of a new, social order. In this scheme, the individual and the whole reinforced each other. Dos Passos, though he espoused a vision of America's potential integration, was part of a group which remained disenfranchised, isolated within a political structure that remained unregenerate:

> all right we are two nations America our nation has been beaten by strangers who have brought the laws and fenced the meadows and cut down the woods for pulp and turned our pleasant cities into slums and sweated the wealth put out of our people and when they want to they have the executioner to throw the switch. . . . on the streets you see only the downcast faces of the beaten nation all the way to the cemetery where the bodies of the immigrants are to be burned we line the curbs in the drizzling rain we crowd the wet sidewalks elbow to elbow silent pale looking with scared eyes at the coffins
>   we stand defeated America.[33]

America, as a corporate entity, did not support either its artists or its workers. Vertov's art grew out of a radical experiment in social integration; Dos Passos's remained the record of a culture unable to transform itself into something whole: "alright we are two nations." Dos Passos's perspective was, nonetheless, a persepctive that Vertov would have called "objective." This identification existed not because of Dos Passos's facticity—although Dos Passos did integrate "authentic" material into his fiction—but because of his breadth of vision, his lack of sentimentality, and his relinquishment of the abstract in favor of the concrete. He had, finally, achieved a view of the whole; he saw the meaning of individual lives within history, the "proper" way to edit and assemble the fragment. Instead of personal alienation, the camera, as Vertov used it, gave Dos Passos a way to imagine social integration both for himself and for the broken world he chronicled in USA. It was Vertov's "interval theory" that allowed him to deal with the problem of surface and underlying structure, for by purposeful juxtaposition, the writer or photographer could use the "surface of positions, orthodoxies, heresies, gossip and the journalistic garbage of the day"[34] to embody the "deep currents of historical change" which lay beneath appearance and beyond the ability of the "simple eye" to perceive.

IV

In *Journeys between Wars,* Dos Passos remembered the time when he had had enough of watching films and gathering impressions of life in Moscow. He went home to work.

> But an onlooker in Moscow is about as out of place as he would be in the assembly line of a Ford plant. If you are an engineer or a mechanic or a schoolteacher you can do something but if you're a

writer you're merely in the unenviable position of standing around and watching other people do the work.

Well, you're a reporter, you tell yourself. You're gathering impressions. What the hell good are impressions? About as valuable as picture postal cards. . . . Worthwhile writing is made of knowledge, feelings that have been trained into the muscles, sights, sounds, tastes, shudders that have been driven into your bones by grim repetition, the modulations of the language you were raised to talk. It's silly to try to report impressions about Moscow; you can stay there and work or you can go home and work.[35]

His memories of 1928 were still framed in terms of the anxiety of looking on without participating, but as he worked on *USA,* that anxiety diminished so that writing became synonymous with work, and work itself the social participation that quieted discomfort. In 1935, explaining his views to the First American Writers' Congress, he still used the language of work to describe his own efforts, calling the writer a "technician just as much as an electrical engineer is." Despite the importance of the analogy itself, despite its usefulness as a way of locating the writer within the culture at large, the description's primary value is in identifying a frame of mind. For Dos Passos does not expand the comparison with industrialized science, but moves instead to discussing the preconditions of all technical work, whether it be engineering or writing. That circumstance is "liberty"; but Dos Passos does not mean simply freedom from police or bureaucratic interference. He means a state of "selfless relaxation" which is "much nearer the way a preacher, propagandist or swivelchair organizer feels. Anybody who has seen war knows the astonishing difference between the attitude of the men at the front, who are killing and dying, and that of the atrocity-haunted citizenry at the rear."[36]

The essential characteristic of both day labor and war, the two terms of Dos Passos's analogy, is not liberty but participation. Liberty or peace of mind is a consequence of participation. The anxiety of being "in the rear," the onlooker, or, to use Dos Passos's own word, the "leech" is what continued to haunt Dos Passos and drove him eventually to war itself. Like his predecessor Stephen Crane, Dos Passos enacted the terms of his own metaphorical preoccupations. Crane's obsession with the American Civil War, which he had never participated in, led him to court danger in the Crimea and in subsequent battles; Dos Passos gave the ultimate proof of his political involvement by organizing and shooting a documentary about the Spanish Civil War (1937). To film the Civil War he had to be in the war; to record danger, he had to face it.

The idea had arisen sometime late in 1936 when Dos Passos, upset by Franco's revolt against the Spanish Republic, searched for ways to influence American opinion and public policy. He wanted especially to induce the Roosevelt administration to allow the Republican government to buy arms in America,[37] and his chosen method, finally, was the cinema: he would participate in making a documentary film about the Spanish people. Nine years after his trip to Russia, he decided to enact the theories of art's embeddedness in history that his Soviet counterparts had encouraged him to follow. If Vertov's theories and film techniques had influenced Dos Passos's narrative prose, Vertov's experience as documentarian of the Revolution influenced his conduct. Dos Passos entered into this venture with a firmly developed sense of the photographer/writer's potential role in history. He emerged from the filmmaking experience with changed ideas. It has been customary to explain Dos Passos's disillusionment in Spain in 1937 as a reaction to the execution of his friend, José Robles, presumably at the hands of the Communists. His search for the truth of the matter was, without question, a highly disturbing and in some ways definitive event in his life. But Dos Passos's retreat from the internal struggles of the Republicans and their supporters was not solely the retreat of a citizen enraged by the duplicity of those he thought to be his allies; for Dos Passos's effort in Spain had been to test a theoretical perspective, to unify art, life, and politics, and it was in all of these areas that Spain proved to be anathema to him. Politics led to the reevaluation of art's role in politics, and in the reevaluation, the work of Joris Ivens, the film's director, played as influential a role as Dos Passos's own investigations into the mysterious death of his friend.

Dos Passos's relation to Ivens is obscured by the later desire of each to be obscure: their political differences, with Dos Passos's swing away from the Communist Party of which Ivens was still a member, led each to suppress stories about their work together; they no longer wanted to be identified as collaborators.[38] But Ivens, as well as Dos Passos, had been schooled by the great documentarians of the Soviet Union; and in his response to Ivens, we can read Dos Passos's further evaluation of the political uses of photography and its meaning to himself as a writer.

When we look back, we can see that Ivens had followed Dos Passos to Moscow in 1929. He came by a different route, and he remained in a more active relation to the film-makers he met there, but it is nonetheless true that both the Dutch cinematographer and the American writer had made similar tours of the Soviet art world at strategic times in their careers. Dos Passos had come as a journalist, gathering impressions and formulating narrative strategies for his first major work; Ivens had been specifically invited by Pudovkin, who recognized the talent of his early experimental films, *The Bridge* (1928), *Rain* (1929), and *Breakers* (1929). But each met and talked with the directors Eisenstein, Pudovkin, and Trauberg, and each had seen the same films. Ivens recalled, "I had seen most of the important Soviet documentary films in Amsterdam and Berlin—*Turksib, Shanghai Document,* the early Vertov films. Now I saw Vertov's newest tour de force: *The Man with the Movie Camera,* and a film called *Spring* by Vertov's brother, Kaufmann ... I also saw Eisenstein's encyclopedia of film technique—*Old and New*.[39]

What Ivens remembered most about his Soviet visit was not the hours he spent in the cinema but the critique of his own work offered by his unschooled Soviet viewers. One incident became emblematic for him, and its retelling in *The Camera and I* (1969) provided him with a way of identifying his own commitment as a photographer/director and a way of conveying his own thoughts about the nature of realism in his work. His audience had seen *Zuiderzee,* his film about the reclamation of farmable coastal land in Holland, and had responded with questions not about the aesthetics of the editing but about Ivens's own status in life. His annoyance gave way to personal satisfaction when he finally understood that the questions were designed to clarify a fundamental confusion about Ivens's class origins; for though he protested that he came from a petit-bourgeois family, the perspective of the film was understood to be that of the working class:

> "You say you are from the middle class, yet the film we have seen was surely made with the eyes of a worker. I know, because it is exactly the way I see work. So either you are a liar and bought the film in Holland from somebody or else you are a worker who's pretending to be from the middle class—and that is certainly not necessary in a workers' and peasant state," [the worker] added smiling.[40]

This has the characteristics of an apocryphal story, but it tells us nonetheless that Ivens understood documentary realism to be a function of the identification of director and subject. It was, according to him, a matter of "declassing" oneself and shedding preconceived views of the subject as "other."

In some ways, Ivens's impulse was the opposite of that of the other great documentarian of the period, John Flaherty, whose films, *Nanook of the North* and *Alamoana,* had succeeded largely through their exoticism, through their demonstration of the foreignness of life in Alaska and Hawaii. Ivens sought, instead, to render a quality of felt life, to make the experience rendered *in* the film the experience *of* the film for those who looked at it. His concern, in reviewing the aesthetic achieve-

ments of his career, was to identify himself, and by extension, his viewers, with great peoples' movements of his time. Consequently, his memories, both in autobiography and interviews, group themselves around
times of participation rather than around analysis; his creativity arose
out of an act of identification rather than out of critique. He was
frankly and satisfactorily a partisan. In *Borinage* (1933), his film about
the aftermath of a Belgian miners' strike, he remembered sharing the
circumstances he filmed:

> No film could have been more concerned with the people than ours.
> Some nights we slept in miners' houses already crowded beyond
> capacity. We not only saw, but experienced things that made the
> Magnitogorsk barracks luxurious by comparison. But the hardships
> and misery became as much a part of our day as it was the miners'.
> We were not strangers. They helped us and worked with us because
> they clearly understood what we wanted of this film—a means of
> bringing their terrible circumstances to the attention of the rest of
> the world.[41]

"Marching," he said, "my camera pretended to be one of the marchers;
the experiment of the *I* film was finally being applied."[42] He referred
in this statement to one of his own earlier experiments, a small work
done in 1929 which had used the camera as the first person and which
had, without sufficient funding, remained unfinished. Like Dos Passos,
Ivens returned from Russia to make his own versions of Vertov's *The
Man with the Movie Camera,* turning the camera on itself, showing the
photographer to be aligned with those who stood before the lens, to be
a worker among peers. In subsequent films, this paradigm remained
intact: to the extent that the camera enacted the circumstances it
recorded, it operated in a realistic mode. Ivens showed none of Vertov's
concern with the symbolic function of the frame, the interstices
between frames, or the editing of these. Nor did he endorse Vertov's
insistence on films that included only unrehearsed events happening
before the witnessing camera. In his eyes, it was perfectly valid to stage
or reenact events in order to deepen the content of the film and even
to "invent events certain to happen in the future."[43] With these views,
he broke entirely from Vertov's beliefs about objectivity and the artist's
relinquishment of personal inclination to become a witness of history.
For Ivens, reenactment was a function of subjectivity, an outgrowth of
the director's sense of the spirit rather than the letter of experience:

> The reenactment introduces ... the integrity of the director—his
> understanding of and approach to reality—his will to tell the essential
> truth about the theme—his comprehension of his responsibility

toward his audience. He is, as an artist, creating a new reality which may influence the thinking of the spectators and stimulate them to action by the truth of his film. No definition of documentary film is complete without these "subjective" factors.[44]

As long as the director's view reinforced or proceeded from the same source as the people who were his subjects, this was not necessarily a problematic position to hold. But Ivens consistently ignored the consequences of disagreement; and though he entered the filming of *The Spanish Earth* with the idealism and aesthetic and political philosophies that characterized his earlier work, disagreement marred its shooting from the start.

The friction over making the film was not anticipated by either Dos Passos or the director. Dos Passos, as a founding member of the Contemporary Historians, had hired Ivens with every reason to believe that they would be compatible colleagues: both could claim analogous schooling in Soviet film; both were deeply concerned about the meaning of the political movements of their time and of the role of the committed artist within those movements. Originally Dos Passos and Archibald MacLeish had asked Ivens's film editor, Helen van Dongen, to assemble a documentary about the Spanish Civil War from existing newsreel footage. When that project proved frustrating—the material they had to work with was primarily taken in support of Franco's revolt—Ivens suggested making a new documentary. Dos Passos and another "well-known American writer" wrote a scenario; they raised money; Ivens wired his cameraman, John Fernhout, and enlisted his support. It was agreed that MacLeish, then editor of *Fortune* magazine, would act as the fundraiser and that Dos Passos would go to Spain.[45] Ivens left New York on 26 December 1936; Dos Passos left on 3 March 1937. Ernest Hemingway, whose presence in Spain so obscured the original plans for the film and the configuration of the film crew, left for Paris on 28 February 1937, three days before Dos Passos.[46]

The events of March and early April are confused by the memories and motives of everyone. Ivens remembers Dos Passos as an early and invaluable team member, someone who helped choose Fuenteduena as the village that would focus the narrative of the film and who acted as translator for himself and Fernhout, who "hadn't even had time to pick up 'camera' Spanish":

> We would sit in the inns and get acquainted with the citizens, the villagers. . . . We had to convince them that we weren't trying to exploit them or capitalize on their war jobs. Fortunately for us, the Alcalde, the most trusted man in the village, showed quick comprehension of what we needed. Dos Passos acted as our translator.[47]

In his memoirs, he then disposed of Dos Passos: he spoke of shooting for four weeks, taking the exposed film to Paris for viewing, meeting Hemingway there and inviting him to join in the film. "That came just right, for Dos Passos had left us in Spain."[48] Hemingway claimed that Dos Passos arrived in Madrid, their base of operation, *after* he did; and in fact, in the later fatuous exchanges that characterized the end of their friendship, Hemingway chastised Dos Passos for not having spent enough time in Spain to have understood the true nature of the conflict. He, of course, did.

All of this is somewhat puzzling, since Dos Passos and Hemingway arrived in Europe within days of each other. Dos Passos could not have spent a month of time with Ivens before Hemingway arrived, though it is possible that he worked with Ivens, returned to Paris and then came down a second time in early April. Only his second trip, if this is the case, is recorded.

Ivens's memoirs must be incorrect. But despite faulty chronology, it does establish that Dos Passos did work with him as originally planned, something Hemingway was anxious to obscure as he pushed himself into an increasingly prominent position on the film. By the time everyone returned to the United States in May, Ivens and Hemingway had established themselves as the only two significant collaborators of *The Spanish Earth:* Ivens edited the footage, Hemingway wrote the final commentary; they showed the film to the Roosevelts in the White House, and they toured it in Hollywood. Neither mentioned Dos Passos. He had simply disappeared.

In fact, he had not disappeared. Josephine Herbst, who was also in Madrid at the time, kept a diary, which she later recognized as sketchy and incomplete—those artists and writers who went to Spain were, after all, subject to the same dangers and stresses that threatened the Spanish population; under seige, few people kept accurate daily records. But we can reconstruct some of the movements of these people. On 16 March Hemingway went from Paris to Barcelona and from there on to Valencia. After traveling among various villages, he arrived in Madrid on 27 March. Dos Passos also traveled among the Spanish villages, and on the eighth of April he wired Ivens from Valencia to announce his imminent arrival. By 10 April, he was in Madrid also. He stayed until after 24 April, when he joined in a radio broadcast to Mecca Temple in New York City. Later in the midst of their deepest enmity, Hemingway liked to speak of Dos Passos as having quit, having deserted ship; but by 26 April, Ivens, too, was in Valencia on his way back to New York City. John Fernhout had been instructed to remain behind to pick up some more shots and to try to find the young soldier whose activities had held the first part of the film together. But by 26

April, the filming of *The Spanish Earth* was essentially over. Dos Passos had stayed to the end.[49]

If we can prove his presence and if we can prove his participation on the film, we have, nonetheless, to explain the animosity that arose before the film team returned from their work. Hemingway, Ivens, and Herbst all remembered Dos Passos's preoccupation during the time with finding news of his missing Spanish friend, Robles. When any motive was attributed to Dos Passos (Ivens explained nothing when he said that Dos Passos had simply "lost interest"), it was that his obsession with Robles precluded any other activities. It was a simple heeding of temperament, a choice. But Robles had been intimately connected with the film; and in fact, it was as an advisor to the scenario that Dos Passos had originally sought him out in Valencia: José Robles Pazos was a professor of Spanish language and culture at the Johns Hopkins University and, at the time, had become a lieutenant colonel in the Spanish Ministry of War: "I knew that with his knowledge and taste he would be the most useful man in Spain for the purposes of our documentary film."[50] The disappearance of Robles, Ivens's attitude toward Dos Passos's distress, coupled with Ivens's working methods, all led to the ultimate sundering of the two. The production of the film itself was the cause of Dos Passos's alienation from it; investigating Robles's death and filming *The Spanish Earth* need not have been inevitably opposed commitments for him.

Ivens went to Spain with a scenario in hand, a working document which he later discarded as the circumstances of war showed him the impossibility of "staging" fictional events in the midst of far-reaching distress:

> How could we ask people who had fought in the fields and in the trenches around Madrid to help reconstruct the atmosphere of King Alfonso's abdication? These people were too deeply involved in their fight to think about how a typical village had behaved before the war. We felt shame at not having realized this. One could not possibly ask people who were engaged in a life and death struggle to be interested in anything outside that struggle.[51]

Ivens remembers interviewing many people as they sought some new line for the script, and he attributes the revised plan to work he did together with Dos Passos:

> It was here that the connection between food and land reform became dramatically clear. We began to work out an approach that would place equal accents on the defense of Madrid and on the small nearby villages linked to the defense because it produced Madrid's food—

potatoes, vegetables, tomatoes, olives . . . people working together for
the common good, the future as well as the present.[52]

The importance of Ivens's memoirs arises out of the admission that he
and Dos Passos agreed on both the original scenario and its revision.
Archibald MacLeish is thought to have helped Dos Passos write the
discarded first plan, but it was a plan that both the artists who actually
went to Spain knew about and condoned. It was not to the previsuali-
zation of the documentary that Dos Passos objected. What, then, did
cause his discontent?

If we look at Ivens's filming procedures—ones in complete accord
with his previous work—I think we can see that he and Dos Passos
were unwittingly on a collision course, that Ivens's enthusiasms and
cinematic techniques would have eventually alienated Dos Passos, even
had Robles not been so ignobly treated and Ivens so unsympathetic
toward his death. For Ivens proceeded to espouse the cause of his subjects, entering into the lives of the villagers and the Fifth Brigade in
the same way that he had worked with the Belgian miners in the Borinage. His fondest self-reflections came from thinking that he lived *in*
his subject, sharing in the dangers and hopes of the Spaniards as if he
were one of them:

> I felt so personally involved in the war, I sometimes lost sight of the
> fact that I had come to do a job and that I would leave Spain when I
> had finished it—whether or not the war had finished. For the first
> time I was really seeing the destruction and death I had read about.

Later he said that his unit "had really become part of the fighting
forces, particularly with the units of the International Brigade."[53]

Hemingway's memories of this time, recorded, in part, in "The Hot
and the Cold," a coda to the commentary for *The Spanish Earth,* bear
this out. Ivens took his camera where the Loyalist troops took their
guns: "Because you had seen a little war when you were young you
knew that Ivens and Fernhout would be killed if they kept on because
they took too many chances."[54]

Such partisan fervor characterized all of Ivens's filming career, and,
indeed, it was this populist enthusiasm tht first led the Contemporary
Historians to solicit his help in their documentary: they knew they all
worked within a particular frame of pro-Loyalist sentiments. Twenty
years later, Dos Passos admitted his own bias when he said, "Our Dutch
director did agree with me that instead of making the film purely a
blood and guts picture we ought to find something being built for the
future amid all the misery and the massacre. We wanted cooperative

work, construction with the profit motive left out to be the theme. We settled on an irrigation scheme being put through by a village collective."⁵⁵ But sympathy with the villages and interest in collective action, which they all shared, were clearly to be distinguished from purposeful distortion. Dos Passos's charge against Ivens was that, in suppressing information about internal discord and repressive measures of the Communist party within the Spanish Loyalist government, he had overstepped the bounds of sympathy. In his opinion, the Loyalists had violated their own principles and had acted in accordance with, rather than in opposition to, the repressive practices of the fascists. This was the meaning of Robles's execution to him: it was analogous to the fates of Sacco and Vanzetti in the United States—who were "each just one man."

> Isn't justice one of the things we are trying to establish? If one honest Spanish patriot had been executed, it's likely that there are more. Our hope was to save the Spanish Republic and all the heritage that went with it. If it had already been destroyed from within what were we fighting for? . . . By the time I left the country there was no doubt in my mind that the case of my friend was no exception.⁵⁶

In his letter to the *New Republic* in 1939, he generalized: Robles's story was one among many, "but it gives us a glimpse into the bloody tangle of ruined lives that underlay the hurray-for-our-side aspects. Understanding the personal histories of a few men, women and children really involved would, I think, free our minds somewhat from the black-is-black and white-is-white obsessions of partisanship."⁵⁷ At a certain point indiscriminate identification with the subject before the lens became a misrepresentation of that subject. To Dos Passos, empathy could become a form of self-delusion, an evasion of truth, a posture of intellectual dishonesty.

Dos Passos's subsequent recoil from the political left and the extreme libertarianism of his later years make him equally subject to the charge of bias. He had, one could argue, simply exchanged one kind of partisanship for another. But his revulsion also tells us that something pivotal happened to him in Spain, and I am interested here only in understanding what that was. For the way Robles's death jarred against the filming of *The Spanish Earth* had implications for Dos Passos's practice of art as well as for his disgust with certain forms of political life.

If we look back on Dos Passos's response to Soviet cinema in the 1920s, we can understand this aspect of his experience more clearly, for what he took from those Russian cinematographers was not simply an enthusiasm for workers and their collective endeavors to shape the future, it was a commitment to a collective view of history, a perspective represented to Vertov, and then later to Dos Passos, by the camera, by the non-human machine, by the viewpoint which no "simple" human eye could achieve. Vertov's manifesto had said:

> Freed from the frame of time and space, I [the camera] coordinate any and all points of time and space, I coordinate any and all points of the universe, wherever I may plot them. My road is toward the creation of a fresh perception of the world. Thus I decipher in a new way the world unknown to you. . . . But it is not enough to show bits of truth on the screen, separate frames of truth. These frames must be thematically organized so that the whole is also truth.[58]

Dos Passos's art was similarly constructed from fragments, from limited perspectives the importance of which became recognizable only through "intervals," through the accumulation of partial perspectives. It was only the whole that contained the meaning of particular historical events and it was only the artist, by espousing "the camera eye," who could see the whole. His integrity consisted in his commitment to this impersonal, collective view. "Freed from the frame of time and space," his task was to collect others' experiences.

It was precisely this commitment to the whole that Ivens's version of *The Spanish Earth* violated. Where Vertov had denigrated individual subjectivity as a limited and thus inevitably false perspective, Ivens gloried in it, elevating the director's subjectivity into the single quality that constituted most a film's realism. Where Vertov had dedicated his life to understanding how editing or montage could embody the reconciliation of multiple perspectives, Ivens had, in Dos Passos's opinion, filmed from a single, fragmented perspective and had elevated that fragment into the whole truth. It was a false claim, and it was to this substitution of a limited point of view for the cumulative image that Dos Passos referred when he castigated Ivens for making propaganda. According to Dos Passos, the truth of the Spanish Civil War could only be revealed in an art form that could include and reconcile disparity. A true film would have to acknowledge and interpret how "honest Spanish patriots" could be murdered in the course of the struggle. What troubled Dos Passos in Spain was not simply that Robles had been killed but that Ivens was engaged in an art that would ignore the meaning of such an event in order to present the war as a clear example of the fight of good against evil. And, indeed, Ivens admitted that he saw himself in these reductionist terms: I [found] myself in Spain, filming a war where ideas of Right and Wrong were clashing in such clear conflict, where fascism was preparing a second world war!"[59]

Dos Passos's final lesson in Spain was one that modified his view of

Vertov as well as of Ivens. For he came to question the possibility of reconciling disparate parts of history into a coherent unity: whatever faith in history had upheld Vertov's aesthetic principles failed for Dos Passos in Spain. Later, looking back on his experience there, he became truculent and self-righteous, as if he alone were holding up the cause of truth and liberty in the face of general barbarism:

> The only weapon we have is telling the truth, or as near as we can get to it. We live in a time when the true history of international events is enormously hard to come by. The truth is not in atrocity stories or partisan propaganda. Lashing ourselves up into a partisan fever will only make us the prey to whatever propaganda the war-mongers want to put over on us. Last time they got us to hate the Huns and the first thing we knew we were persecuting the Wobblies.[60]

When he left Spain in that spring of 1937, however, this ideological position had not yet ossified. Instead, he knew only that he could not see the whole picture himself: others might deceive themselves in ways he could identify, but he did not claim to hold an alternate truth. On his way out of Madrid, gathering materials for the essays that would constitute his alternative to *The Spanish Earth,* he felt increasingly depressed. He had come to Spain with a love of the villages and with a profound respect for the common, undistinguished men and women of the countryside. He could see that for them to win against the fas-cists, they would need to create some powerful "industrial organization ... within anarcho-syndicalist ideas."[61] But the villages were not united; they were not organized in the face of such enormous opposition. In the final essay of *Journeys between Wars,* Dos Passos recalled a man from Barcelona saying, "It's the villages.... They want to know what to do." Dos Passos's response was a further pessimism:

> How can they win, I was thinking? How can the new world full of confusion and cross purposes and illusions and dazzled by the mirage of idealistic phrases win against the iron combination of men accus-tomed to run things who have only one idea binding them together to hold on to what they've got?[62]

He did not see people working together as Ivens wanted to show them, in a heroic documentary about the little man contributing his part to an overwhelming good—the fight against fascism. For Dos Passos, the fragments did not cohere: Robles's death had been a fragment that could not be edited nicely into place; it could not be identified as a detail that contributed to a larger coherence. Nor did Dos Passos under-stand the villages of Spain to be coherent, united in a firm vision of their own collective future: "They want to know what to do."

USA, for all of the decadence recorded in the lives of its characters, had been created out of optimism and hope that clashing opposites were the precursors of a larger, historical synthesis. Its broken world was a world capable of mending. *Journeys between Wars* was the record of a changed consciousness: it ended with the disquieting image of a failed radio broadcast. In the same way that he had seen the emblematic qualities of Vertov's "The Cinema Eye," the mechanical tool that gathered up and united disparate parts, so Dos Passos understood this radio broadcast to have metaphoric implications: it was a failed communication between two points. To him, the two Spanish cities were analogous to two film frames; they were isolated points whose coherence depended on the editor, the assembler of parts, the maker of a larger purpose, whether that maker be History in its great sweep toward eventual socialist organization, or the artist who stood as a revealer of that historical purpose. But this time, the world remained in fragments for Dos Passos: he could not find its underlying coherence; he could not hear the broadcast between the two points. The Spanish cities remained isolated. His book—and this stage of his political life—ended in "sharp crashes of static."[63]

Joris Ivens. Film stills from *The Spanish Earth (IMP/GEH Still Collection)*

Dziga Vertov. Film stills from *The Man with the Movie Camera*. 1928. Reproduced from the book *Constructivism in Film: "The Man with the Movie Camera"* by Vlada Petric (Cambridge: 1987). Courtesy of Cambridge University Press.

Walker Evans. Dora Mae Tengle [Evans's shadow in foreground]. Hale County, Alabama, 1936 *(Library of Congress)*

5

James Agee and Walker Evans

Representing Poverty during the Great Depression

It seems to hang suspended in the middle distance
forever and forever.—William Faulkner

I

IN his thirties, James Agee became one of the major film critics of
his generation. That particular talent may not have been apparent
in the summer of 1936 when he and Walker Evans drove down to
Alabama together, but even then Agee was clearly tuned to the importance of his friend's still photography. In fact, we can think of Agee as
the writer who, with excruciating self-reflectiveness, made explicit the
issues of seeing and power, of dispossession and cultural definition, that
had been the more covert concern of his artistic precursors. Certainly
he knew that Evans was the only photographer he wanted to work with
on his *Fortune* magazine article about cotton tenancy, and he had pushed
around two formidable institutions to arrange the joint assignment. *Fortune* had been persuaded not to send one of its regular photographers to
Alabama, and the Farm Security Administration—Evans's employer—
had been asked to release a member of an already small staff. The agreement was a compromise for everyone except the two young men: *Fortune* would pay Evans's salary and expenses; the government would own
the negatives from the trip.

It is no longer possible to know what kind of rapport existed between the two during the unusual circumstances that characterized their two months in the South, but in 1960, Evans still remembered his friend vividly: as a rule, Agee had been embarrassing, exuberant, indelicate, and angry. He remembered how they both had lived in the home of one of the farmers, and how Agee had suppressed his native rebelliousness to be casual and friendly and easy with them. "In one way," Evans remarked, "conditions there were ideal. He could live inside the subject, with no distractions." Evans did not mention the terrible conflict that ensued when Agee returned north and refused to hand in a concise, informative, and self-contained essay about cotton farming that his editors at *Fortune* had expected. But he did believe in the fruits of his friend's struggle. To him, Agee's talent was "unquenchable, self-damaging, deeply principled, infinitely costly, and ultimately priceless."[1]

Evans's short memoir, carefully, beautifully written, is the tribute a man writes when his own life has been deeply touched by someone else's conduct. It is, in the manner of his photographs, both sparse and evocative; and like his camera work, it is also scrupulously self-effacing. Never does he speak of himself in relation to Agee, yet he must have known that in the matter of influence, he had held equal sway, that he, the quieter and less explosive of the two men, had had many things to teach. If Evans did not call attention to his own creative influence, Agee fortunately did; and the records of those lessons are dispersed throughout the text of *Let Us Now Praise Famous Men*. In fact, Agee's struggle to come to terms with the camera forms a coherent and powerful subtext, an underground commentary, making this book a central document—perhaps *the* central document—of art's struggle with social responsibility during the Depression.

What Agee wanted to avoid was clear to him. He was, as Evans notes, "in flight from New York magazine editorial offices, from Greenwich Village social-intellectual evenings,"[2] from misplaced middle-class prurience; and he admitted it candidly: "It seems to me curious," Agee wrote, "not to say obscene and thoroughly terrifying, that it could occur to . . . an organ of journalism, to pry intimately into the lives of an undefended and appallingly damaged group of human beings . . . for the purpose of parading the nakedness, disadvantage and humiliation of these lives before another group of human beings, in the name of science."[3]

He could also identify his own alternative commitments. He wanted to find a humane way to live and write, a way to gather material that did not damage those observed, a way, in short, to balance the moral claims of the tenant farmers against his own desire to produce a text about them. "The effort," he wrote, "is to recognize the stature of a portion of unimagined existence, and to contrive techniques proper to its recording, communication, analysis, and defense" (11). "For in the immediate world," he claimed, "everything is to be discerned . . . centrally and simply, without dissection into science, or digestion into art" (11). But how he could achieve these goals was less clear to him. It was here, in the matter of proper technique, that Walker Evans provided guidance; for in watching Evans work, Agee acquired a way of understanding the problems and opportunities of his own unprecedented text. "Next to unassisted and weaponless consciousness," he said, "[the camera is] the central instrument of our time" (11).

With his own narrative goals in mind, he began to sort out the camera's use and misuse and to clarify his own relation to the medium. This was so pressing an issue that he framed his own text with comments about the subject. He began his book by telling how he had helped Evans take pictures of a church when they first arrived in the South, and ended it with a critique of Margaret Bourke-White's photography book, *You Have Seen Their Faces.* Superficially these anecdotes have little to do with the subject of farming conditions in Alabama, but they are nonetheless essential to Agee's struggle to redefine an art that could render those conditions responsibly. Placed strategically at the beginning and end of Agee's own prose, these vignettes can be seen as items in a private play of imagination, as Agee's symbols of an integrity sought and a danger escaped. Through these photographers, techniques of vision became objects of vision. By setting up cameras, taking positions, and relating to the subjects of their photographs, they provided Agee with visible tableaus of the more intangible dilemma he faced as a writer. Just as Evans stood as an example of all that Agee hoped to achieve, Bourke-White represented all that he hated about investigative reporting and the New York art world.

II

As Agee envisioned her, Margaret Bourke-White was a flighty, self-absorbed woman. In the final pages of *Famous Men,* he tried to diminish her by quoting an interview from a women's magazine: "Margaret Bourke-White Finds/Plenty of Time to Enjoy Life/Along with her Camera Work" (450). He scorned her private pastimes, her concern for expensive clothing, her cagey and obviously competent career management. "She's a tango expert; crazy about the theatre, loves swimming, iceskating, skiing, and adores horseback riding. Sometimes, she explained, when she knows that the light will be right only a few hours of the day for whatever pictures she is taking, she has her horse brought around to 'location' and rides until the light is right" (454). He was angry that she was one of the highest paid women in the country with the resources to indulge her whims. At the time Agee knew her, Bourke-White had worked for the federal government, for private businesses, and for his own employer, *Fortune* magazine. In fact, she was widely and popularly known as the woman Henry Luce had chosen as the first photographer for *Fortune* and had chosen again as his lead photographer for *Life*. She also had just published a well-received and influential book on southern sharecroppers, *You Have Seen Their Faces* (1937). Like Agee's own book, it contained both pictures and text; Bourke-White had worked with (and later married) the writer Erskine Caldwell. Looking back with the evaluative stance that the passage of time allows, we can see that Bourke-White was largely responsible for the convergence of photography and journalism in this country and that she was driven by faith in the richness and veracity of the photo essay—the story that is built from cumulative images.

From her own point of view, Bourke-White's trip south had marked a turning point in her life that was analogous to Dorothea Lange's decision to leave studio photography and take her camera down into the breadlines of San Francisco. Both young women were talented and successful, and both had apparently felt the discrepancy between their own safe fortunes and those of the people left destitute by the Depression. Each responded to her personal crisis by finding new, more socially relevant materials to photograph. Bourke-White had to close lucrative advertising accounts and temporarily drop off the payrolls of *Fortune* and *Life*, in order to find time to devote to this work. Though both Agee and Evans felt bitter about the money Bourke-White made from her book, financial gain had not been her motive. In fact, her desires had been of the best. She and Caldwell had wanted quite sincerely to alleviate suffering, and they both believed that information of the sort the camera and first-person reporting could provide was essential to this endeavor. To call attention to the economic weakness of southern agriculture and to the degradation of sharecropping was, in their eyes, tantamount to providing a solution. It was, at the least, a contribution that individuals like themselves could make. Beyond recommending further government study, Caldwell's text did not advocate specific programs of action, but the assumption behind the effort of these two artists was clearly that decent citizens could not remain passive in the face ("you have seen their faces") of social wrong.

Although Bourke-White remembered this book and this trip as a turning point in her career, we can see in retrospect that her own thoughts about the camera and its mission in the South were a reiteration of the philosophy that also guided her customary work for *Life* magazine. "I believe," she said, "that photographs are a true interpretation. One photograph might lie, but a group of pictures can't. I could have taken one picture of share-croppers, for example, showing them toasting their toes and playing their banjos and being pretty happy. In a group of pictures, however, you would have seen the cracks on the wall and the expressions of their faces. In the last analysis, photographs really have to tell the truth; the sum total is a true interpretation. Whatever facts a person writes have to be colored by his prejudice and bias. With a camera, the shutter opens and it closes and the only rays that come in to be registered come directly from the object in front of you" (454). In fact, I would argue that this perspective underlies and explains the work of Bourke-White's entire career, and it is this circumstance that lends Agee's criticism of her work during the Depression especial weight: to him *You Have Seen Their Faces* was emblematic of far more than itself. We might say that Bourke-White's photographs and her method of taking them had simply provided Agee with the occasion to express a much more general indignation about art's unwitting betrayal of its own power. Even though the book had not been commissioned and despite the fact it had no sponsoring agency, Agee thought that Bourke-White's simplistic formulations about truth led her to misperceive the nature of the cultural transaction in which she participated. In picking on her, Agee was criticizing a mode of class interaction, a set of attitudes, and a kind of blindness that were much broader than Bourke-White's personal experience.

It is only in this way that we can understand the vehemence of his attack, for he tore the book apart, beginning with the characteristics that formalist critics would most admire: he pointed out that Bourke-White was noted or the drama of her shots, for strained angles, for obvious arrangements of items and people into satisfying compositions. To Bourke-White, these maneveurs were the staples of her trade, and her success in mastering them signalled both her professionalism and her artistry. She loved form, and at least at the beginning of her career, she regarded human life insofar as it could be abstracted and patterned in a print. In her autobiography she distinctly remembered the occasion when she first realized that human beings could be an interesting subject for the lens: she had been photographing the Dust Bowl in 1934. Though her photographs from this assignment are mostly the records of soil erosion, dead crops, and abandoned houses, she must have realized the human misery this desolate landscape implied. In 1936, she remembered, "This was the beginning of my awareness of people in a human, sympathetic sense as subjects for the camera and photographed against a wider canvas than I had perceived before."[4] She learned, as she later said, "a man is more than a figure to put into the background for scale."[5]

To both Agee and Evans these attitudes and photographic strategies were signs of deplorable moral ignorance. Agee thought the book was an example of journalism's own complacent delusion that it is "telling the truth" (234). As he saw it, *You Have See Their Faces* was propaganda which in New York was not recognized to be so because the attitudes it displayed were so widely shared. Malcolm Cowley had commented that the "quotations printed beneath the photographs are exactly right; the photographs themselves are almost beyond praise,"[6] but Agee found the book sensational, condescending, and brutal, and he argued this from both the evidence of the photographs, and his knowledge of her working methods. Bourke-White's postscript to the book, "Notes on Photographs," provided him with the ammunition he most needed, for it began with a noteworthy admission: "When we first discussed plans for *You Have Seen Their Faces,* the first thought was of lighting."[7] Agee took this to be emblematic of an entire set of misplaced commitments. He felt that Bourke-White considered the photograph, the art object, to be more important than was the subject in front of the camera, and he found numerous examples of her insensitive priorities:

> This is the young lady who spent months of her own time in the last two years traveling the backroads of the deep South bribing, cajoling, and sometimes browbeating her way in to photograph Negroes, share-croppers and tenant farmers in their own environments (451).

He pointed out that Bourke-White worked as if she were in collusion with her middle-class viewers and that even her own words admitted a desire for effect at the expense of immediate human transactions. She had, after all, been satisfied with the photographs of a Negro minister whose colorful oratory was captured by climbing, unbidden, into his church: "I believe the only reason we were successful was because the minister had never had such a situation to meet before. Photographers walking into the middle of a sermon and shooting off flash bulbs were something he had never had to contend with."[8] Agee's point was that these photographic subjects were being used and that they were exploited in the very act that purported to help them. Had he focused his attention on only the momentary discomfort of the people in front of the camera, his reaction might be considered hypersensitive, but he

saw the transaction in a much larger context. His most devastating criticism was that Bourke-White had captured the face of nothing real; in
particular, her photographs had been dangerously narcissistic.

In leveling this charge, Agee could not have known Bourke-
White's memory of a lesson from this trip, for it was not published
until many years later:

> We went into the cabin to photograph a Negro woman. . . . She had
> a bureau made of a wooden box with a curtain tacked to it and lots
> of little homemade things. I rearranged everything. After we left,
> Erskine spoke to me about it. How neat her bureau had been. How
> she must have valued all her little possessions and how she had them
> tidily arranged *her way,* which was not my way. This was a new point
> of view to me. I felt I had done violence.[9]

But he could tell from "Notes on Photographs" that Bourke-White's
procedures generally were more extended versions of this same kind of
purposeful rearrangement:

> Sometimes I would set up the camera in a corner of the room, sit
> some distance away from it with a remote control in my hand, and
> watch our people while Mr. Caldwell talked with them. It might be
> an hour before their faces or gestures gave us what we were trying to
> express, but the instant it occurred the scene was imprisoned on a
> sheet of film before they knew what had happened.[10]

The key phrase was "their faces or gestures gave us what we were
trying to express," for it told Agee that Bourke-White had consistently
photographed her own preconceptions of poverty. Apparently, the book
carried a disturbing message to others beside Agee. In 1939, the Southern Combed Yarn Spinners Association published its own photographic
essay, an essay intended from its title—*Faces We See*—to stand as a
direct rebuke to the degradation with which Bourke-White so insis-
tently characterized the working people of the South. It too can be said
to serve a propagandistic purpose, for it counters Bourke-White's condescension with an unremittingly wholesome picture of the workers'
lives, substituting boy scouts, cheerleaders, and football teams for
Bourke-White's barefooted and lazy fishermen; quiet churches for the
disproportionate zeal of her gospel preachers; neat houses with refrig-
erators, indoor plumbing, and electricity for the crude shacks that lit-
tered her southern landscape. "These people lead normal American
lives," says Mildred Gwin Barnwell, "[and they] wonder what part of
the South folks are talking about when it's called Economic Problem
No. 1."[11]

While *Faces We See* sets a pattern that is too adamant about the
steady employment, good wages, and pleasant community life of mill
workers, while it portrays the mill as a benevolent mother "throbbing
and humming with the sustenance it [gives] to lives dependent upon it
for critter comforts," it is nonetheless true that it articulates a sense of
self-worth or dignity that remained outside of Bourke-White's ability
to see. It is as if the book were created to remind Bourke-White that
the insider's view, the view engendered by familiarity and trust, was
the *sine qua non* of truth, created to make her see her photographic tourism for the false vernacularism it was.

Agee would have recognized the self-interest that generated *Faces
We See;* he would have distrusted the book's insistence that "in the mill-
villages evidence of well-being and contentment is reflected in the faces
we see," and he would have been angry that it misused the photograph's
claim to veracity ("seeing is believing, you know, and the camera
doesn't lie"[12]) in the interests of another partisan view. But these res-
ervations would not have weakened his specific criticism of *You Have
Seen Their Faces.*

Behind Agee's mockery was a growing awareness that Bourke-
White was dangerous *not* because she was uniquely silly but because she
played into a network of rules of coherence and strategies of explanation that characterized the communications industry as a whole. To
Agee, Bourke-White epitomized the policies of reporting of *Time, Life,*
and *Fortune.* These in turn represented something more difficult to identify but nonetheless important as a national attitude. The seriousness of
Agee's charge against Bourke-White derives from his association of her
photographic interference with larger, more diffuse forms of social control. Through her he understood that institutions around him, including those that employed photographers like Bourke-White, were
involved in creating self-serving concepts of poverty and that it was
their own interest in the status quo that gave them incentive to represent the poor as worn, repugnant, alien, and stupid. Those who were
helped by visual information of the sort Bourke-White produced,
according to Agee, were those for whom the photographs were taken,
not those whose lives stood so poorly exposed. Whose faces have we *not*
seen is usually an important question to ask, for a favored mode of
power is to be inscrutable.

Agee's every effort in writing *Let Us Now Praise Famous Men* was to
avoid unwitting participation in the indirect damage or, rather, the
ultimate neglect or indifference of such complacent art. Although this
commitment stemmed partly from Agee's confusion of photographs
with the use to which they can be put, he wanted to ensure that no one

could use his text for tactical convenience. He would write something that would lend itself to no one's reading comfort. This determination can explain both the insults and self-conscious nastiness of the book and Agee's refusal to advocate any form of social program for alleviating the sharecropping situation. I think that he agreed with Bourke-White and Caldwell's belief that "seeing" had to be the prerequisite for social action. But if Bourke-White, the most highly acclaimed "seer" of her time, could not see the actual parameters of life in the rural South, then how could any viable suggestions for change derive from her work?

Because Agee saw the extent and possible harm of her condescension, he himself remained committed to "pure" presentation insofar as that was possible, and to self-presentation when he understood the nature of his own interference. Walker Evans knew this about his friend: "You notice," he recalled, "that Agee is saying ad nauseum almost throughout the book: 'For God's sake, we must *not* exploit these people, and how awful it is if we are. . . . ' You didn't find that in Bourke-White anywhere. Not even awareness of the fact that she should have felt this."[13] Agee called his posture "the effort to perceive simply the cruel radiance of what is." In effect this effort was an attempt to avoid class-related preconceptions of poverty, and it involved him equally in self-forgetfulness and in self-consciousness. As he turned from Bourke-White in anger, he looked to Evans for guidance and to the camera's proper use as a model for his own narration.

III

The reasons for Agee's intense respect for Evans are not hard to discover. He thought his friend to be unique and placed him next to Christ and Blake as one of the moral elite whose sensibilities had risen above the assumptions of their own limited cultures. In a poem that prefaces the main text of *Famous Men*, Agee called Evans a spy "moving delicately among the enemy," and summoned him to "order the facade of the listless summer." Whether Christ, Blake, and Evans would have proved good companions is quite beyond me, but Evans did, at least, have some clear ideas about using his camera.

Perhaps the broadest way of locating Evans among his contemporaries would be in a context of rebellion that is expressed by a dream Edward Weston had while in Mexico in the 1920s. Weston woke from sleep repeating, "Alfred Stieglitz is dead. Alfred Stieglitz is dead," a refrain that occasioned some painful self-reflection. "The obvious way to interpret the dream," he finally concluded, "would be in the forecast of a radical change in my photographic viewpoint, a gradual dying of my present attitude, for Stieglitz has most assuredly been a symbol for an ideal in photography towards which I have worked in recent years."[14]

Rather naturally, Weston interpreted his dream in terms of his own situation, but he might also have spoken for a whole generation of photographers who lived in Stieglitz's shadow and whose creative independence demanded the master's "death." To anyone who looks at the evidence of Stieglitz's career, the variety and dynamism of his photography is obvious. But despite his catholicism, his explorations, and his eventual commitment to make precise, clean photographic images, he remained a symbol for one, limited "ideal in photography." To younger photographers, including Evans and many of his peers, Stieglitz represented pictorialism, the salon world, life at one privileged remove from the ordinary; and they shaped their own sensibilities in response to him.

As a young photographer, Evans had gone to see Stieglitz, just as Weston had done before him, and had received scant attention. Weston came away from the 291 salon with a glow that was later dulled by experience and which he finally repudiated, but Evans disdained Stieglitz from first to last:

> When I started photographing in 1929 I was working against what I considered the dishonesties of the camera in the hands of the two reigning masters of the time: Steichen and Stieglitz. I thought Steichen was too commercial and Stieglitz too arty, playing around, photographing the beautiful, calling it "God"—I thought it was nonsense. I was working from anger; I was furious at the old boy. I wanted rather violently to tell *them* the truth—or anybody who was used to looking at swans and reflections in a pool, the whole pictorial school, the "salon" photographers. And that was a very good stimulus, gave me some pride.[15]

Were these animosities only the grudging responses of unacknowledged artists to those in popular favor, they would not bear repeating. But both Weston and Evans were articulating a widespread dissatisfaction, and their complaints, as well as their achievements, can serve as an index of a belief shared among many photographers that they should turn away from photographic manipulations and instead play a less intrusive role in setting up scenes and in achieving special effects through printing procedures.

Paul Strand, Edward Weston, Ansel Adams, some of the other photographers in Group f64, and Walker Evans—all of them unique in

choice of subject and in technical matters—shared a common intoler-
ance for the conventions of photographic idealism and a common commitment to rendering exact, unmediated experience insofar as such
transparency was possible to them. They wanted to bridge the gap
between photographs taken with artistic intent and those taken for utilitarian purposes. They considered this commitment to entail a reliance
on "the basic properties of the camera, lens and emulsion"; that is, they
intended to avoid special effects, retouching, and imitation of the other
arts and instead to offer their subjects "as revealed by the natural play
of light and shade ... without disguise or attempt at interpretation."[16]
Where a previous generation of photographers had tended to use soft
focus and strong contrasts of light and shadow, Evans and his contemporaries chose to emphasize the camera's unique capacity for impecca-
ble resolution and rendering of detail. Paul Strand represented common
sentiments when he explained:

> The photographer's problem is to see clearly the limitations and at
> the same time the potential qualities of his medium, for it is precisely
> here that honesty no less than intensity of vision is the prerequisite
> of a living expression. This means a real respect for the thing in front
> of him expressed ... through a range of almost infinite tonal values
> which lie beyond the skill of the human hand. The fullest realization
> of this is accomplished without tricks of process or manipulation
> through the use of straight photographic methods.[17]

"Straight photographic methods" were certainly Evans's goals as well.
"In documentary," he commented, "not only is actuality untouched by
the recorder as much as may be, it is unInfluenced. . . . The documentary artist ... does what he can not to change it spiritually. He tries to
add nothing to it: no ideology, no polemic, no extrinsic excitement, no
razzamatazz technique."[18]

These principles underlie and explain the now familiar stories of
Evans's disregard for his employer, Roy Stryker, at the Farm Security
Administration. The agency had a clear function: to provide photographic evidence of rural poverty and to give later evidence that various New Deal economic policies had successfully alleviated the problems originally documented. In Stryker's eyes, photography had an
unassailable social use; but in Evans's opinion, any overt use of photographic images was a violation of both the photographer and the world
he purported to represent: he lived in a world where utility and integrity were antithetical values.

Stieglitz had seen integrity in terms of aesthetics: to crop a picture
was taboo; to photograph something in the wrong place was a flaw in

composition. Although Evans, too, was masterful at seeing the formal qualities of life within the visual field of the camera, he defined integrity in terms of fidelity to life itself. To him the claims of honest recording preceded the beauties of formal arrangement and preceded any preordained use by a sponsoring agency. Margaret Bourke-White was, quite understandably, anathema to him. Her tampering, as much as anything Stieglitz or Steichen did in the name of art, provided the grounds of general distrust of the medium, whereas he wanted people to rely on his photographs as unassailable and authoritative revelations of human life. He saw his role as an essentially self-forgetful one, and, though he knew that photographers could evoke the conditions they photographed and could imply, by virtue of focus, distance, and camera angle, an attitude toward them, he himself would intrude as little as humanly possible. He would act as the agent of other people's experience, much as his mid-nineteenth century predecessors had, as a conduit to realities not present except through the pictures themselves.

Evans's rejection of ideological motivations, his insistence on his own autonomy, can be seen, paradoxically, as the key to his own political influence. "I do have a weakness for the disadvantaged," he admitted, "but I'm suspicious of it. I have to be, because that should not be the motive for artistic or aesthetic action. If it is, your work is either sentimental or motivated toward 'improving society,' let us say."[19] Detached, fastidious, trying to evade interested political perspectives, he gained a following and acquired his own authority. His concern was to give the nation images that revealed itself to itself as it was—not as the New Deal government, the middle class or the business world wanted it to appear. This attitude is, of course, a highly problematic one. For one thing, it is easily argued that there is no such thing as a non-partisan or disinterested point of view. In choosing the stance that so frequently characterized his photographs—a middle distance, the camera held at eye level with no special lighting effects—Evans was making aesthetic choices that were just as noticeable as, for example, Stieglitz's decision to hold the camera at close range: "I have put my lens a foot from the sitter's face because I thought when talking intimately one doesn't stand ten feet away"[20] (presumably referring to a photograph of Georgia O'Keeffe). Both men were using the physical distance of the photographic moment as a metaphor for psychic distance: closeness implied intimacy to Stieglitz; medium distance—perhaps because it avoided extremes—provided Evans with an aesthetic vocabulary for neutrality or respect. But one should also be aware that the aesthetics of neutrality is itself a problematic concept, for documentary is inevitably a way of organizing and expressing social knowledge. Vision is never neutral; the transference of knowledge is not any more innocent than is the transference of power, and this is so whether a single artist or a broadly influential agency sponsors the documentary material. In trying to find an aesthetic escape from the clutches of established structures of authority, Evans was placing an extraordinary burden on but giving an extraordinary opportunity to the individual photographer. By elevating the authority of private sensibility over that of the country's whole bureaucratic structure, he was redefining the artist's role within the larger cultural and political world. Should we look at Evans as modest and self-effacing in his respect for his subjects? Should we look at him as a provocateur in relation to his audience? Should we see both attitudes reconciled in a larger certainty of purpose? I am not sure how many of these complicated issues Agee saw. Evans came into his life at a time when he was already struggling with the formal problems of shaping a set of rejections into another acceptable form of narrative art. Agee found in his friend what he needed to find: Evans's methods of disinterest were what most profoundly interested him.

The relationship of the two men was probably a complicated one, but after noticing that Agee writes about it consistently in *Famous Men,* we can return to the written text and read it, at least in part, as Agee's fictional reenactment of Evans's photographic techniques. In fact, one could almost say that Agee began to see himself in Walker Evans and to regard the camera in the same way that Henry James sometimes regarded his fictional narrators: as "a convenient substitute ... for the creative power otherwise so veiled and disembodied."[21]

Although it is difficult to trace the specific origin of Agee's private equation of camera and pen, it did exist that summer to the extent that Evans photographing a small, southern church, Evans setting up a tripod in front of the Ricketts's farmhouse, and Evans sneaking shots of people unawares became analogues of Agee's own position in Alabama. Perhaps the fact that Evans carried and used a trunk full of equipment made it easier for Agee to see the nature of the transaction they both confronted. Others could say to Evans, "Sure, of course, take all the snaps you're a mind to" (25). No one noticed the acuity of Agee's native perception; he had nothing tangible to identify him as an artist and could "pass" in ways that Evans, with his cameras, could not. If Agee's talents were "veiled and disembodied," Evans's skills were not; Agee wrote about his friend's photography, emulated it, and was only able to extricate himself from the camera's influence by understanding the limits as well as the particular capabilities of photography.

Agee was principally impressed by the details that the camera could scrupulously record. "One reason I so deeply care for the camera," he said, "is just this. So far as it goes . . . and handled clearly and literally in its own terms, as an ice-cold some ways limited, some ways more capable eye, it is, like the phonograph record and like scientific instruments and unlike other leverages of art, incapable of recording anything but absolute, dry truth" (233). If Margaret Bourke-White had distorted and manipulated her prints, Evans had, in Agee's opinion, handled the camera "cleanly and literally in its own terms," and this is what he, too, would do with language.

Agee's most obvious imitations of Evans were rather naïve translations of photographic panning shots into "equivalent" prose. In the same way that a moving camera could scan a room, taking in detail after detail, Agee would offer up indiscriminate description. On one afternoon when the Gudger family had gone visiting, he snooped through all their possessions and recorded what he saw, even down to the dust:

> And here a moving camera might know of its bareness, the standing of the four iron feet of a bed, the wood of a chair, the scrolled treadle of a sewing machine, the standing up at right angles of plain wood out of plain wood, the great and handsome grains and scars of this vertical and prostrate wood (149).

This is one small example of the heavy encyclopedic descriptions with which Agee repeatedly burdened his narrative—all of these passages literal responses to photographic techniques. But Agee extended these details as an act of reverence, as a way of paying homage to the superior artistry of the existing world: "These plainest and most casual actions are . . . more beautiful and more valuable, I feel, than, say, the sonnet form." For someone of Agee's broadly religious inclinations, this respect might have been extended to anything or anyone. But the disparity between himself and the Gudger family made him all the more aware of the privilege of perception and of the responsibilities entailed in exposing their lives. Like Hawthorne before him, he talked about being "privileged by stealth to behold" what was not rightfully his. Had he carried self-reflection into the imaginative New England past, he might have seen himself as Holgrave in *The House of the Seven Gables* or as Coverdale at Blithedale Farm—living among people whose life experiences were valued with ulterior motives; all of them, as Hawthorne had said, "turning the affair into a ballad." In positing the persistent theme of problematic voyeurism, the distinctions between the two artists are as crucial as their similarity, for Hawthorne tended to associate his doubts about secret observation with the quirks of his own psyche and to see them as a private anxiety, whereas Agee related his guilt to an entire cultural milieu and to a series of strategies encouraged by *Fortune* magazine. He brought private anxiety into corporate relevance, associating his own required voyeurism with the manipulations and interests of an entire culture. Unlike Hawthorne, who expressed his anxiety and then remained caught in his own dilemma, Agee struggled to find some kind of solution to the seemingly antithetical demands of community and observation.

Put simply, he required of himself to be a responsible observer (according to his own definition of responsibility) and to stand exposed to the degree that he asked others to bare themselves to him. If his interminable lists of details were naïve imitations of Evans's camera, they were at least deeply premeditated, and offered as an example of responsibility and as an alternative to the condescending selections he had regretted in Bourke-White's work. Nothing would be unworthy of his attention and nothing would be invented.

To this end, Agee continued to posture as a camera, wishing, as Hawthorne had also wished, to become a disembodied eye, to be someone who could be present without affecting the situation observed. Even Agee's initial request to the Gudgers was framed in these terms. He wanted to ask "whether among you all . . . we might live, paying our room and board, but with nothing at all changed because we were there" (370–71). But even as he asked this, he knew that he would be living through a situation of unnatural or at least unwonted intimacy with a group of strangers whose lives he might profoundly influence. "But it is a cruel and ridiculous and restricted situation," he admitted, "and everyone to some extent realizes it" (62).

The text reveals that he balanced the tension between participating and recording first by writing passages of "pure" perception, as if he were a lens or the sensitive plate of a camera, and then by extricating himself from that imaginative role and dramatizing himself in the act of perceiving and interacting. On one hand, Agee was the camera itself, registering impressions, making direct and unmediated contact with the world; on the other hand, he was Walker Evans, the holder of the camera, a cameraman who inevitably had to act as a man in relation to others. In pursuit of his first goal, he would not only describe what he saw, but describe from a clearly located point of view. Of the Gudgers' house he wrote: "We stand first facing it, squarely in front of it, in the huge and peaceful light of this August morning." And as he continued writing, he realized that for the camera, registration occurred in relation to the quality of light during exposure, and so he noted: "In full symmetry of the sun, the surfaces are dazzling silver, the shadows strong as knives and india ink, yet the grain and all detail clear: in slanted light, all slantings and sharpening of shadow: in smothered light, the aspect of bone, a relic: at night" (146) it is different.

The bulk of Agee's "camera-acting" occurred as he persisted in concretizing his experience, as he struggled to convey the "materials, forms, colors, bulk, textures, space relations, shapes of light and shade" of his life in Alabama. But the extent of his preoccupation with photography was revealed even in his metaphors: the dawn emerged "like a print in a tank" (87). Looking at the window in the Ricketts's kitchen, Agee noticed that the "brightness though powerful is restricted, fragile and chemical like that of a flash bulb" (198). Most significantly, when he backed off and tried to identify the ignorance of the tenant farmers, their special isolation and narrowness, he did so by contrasting their simple awareness with his own "agonies of perception." As he struggled with the burdens of consciousness, his own world became filled with a "swarm and slime of monsters." But in the sharecropper's world, "few such beasts exist, and the instruments whereby he might see them if they did . . . the lenses of these are smashed in his infancy; the adjustment screws are blocked; his is a more nearly purely a tactile, a fragrant, a visible, physical world" (108).

Despite the incidental photographic metaphors which he used to identify the tenants, Agee considered the camera a tool of educated and developed consciousness. That no tenant farmer could buy a camera was a sign of more than poverty—it indicated the full extent of their bondage, the weight of a world restricted for want of a vantage point which might render it explicable and thus subject to control. To take a picture was to have a perspective that no sharecropper could claim. Instead the Gudgers, the Rickettses, the Woodses took the only role open to them: that of subjects for the camera, objects of interest for more privileged eyes.

Of this Agee was painfully aware, and he wrote about Evans's photographic sessions both because they happened and because he knew that he was as much a spy as Evans. He remembered:

> [T]here on one side of the porch of the house, Walker made pictures, with the big camera; all to stand there on the porch as you were in the average sorrow of your working dirt and get your pictures made; and to you [Mrs. Ricketts] it was as if you and your children and your husband and these others were stood there naked in front of the cold absorption of the camera in all your shame and pitiableness to be pried into and laughed at; and your eyes were wild with fury and shame and fear. . . . [A]nd Walker setting up the terrible structure of the tripod crested by the black square heavy head, dangerous as that of a hunchback, of the camera; stopping beneath cloak and cloud of

wicked cloth, and twisting buttons; a witchcraft preparing, colder than keenest ice, and incalculaby cruel (363-64).

Agee attributed shame and terror to the farming families, and he emphasized their naïveté, their inability even to recognize that candid shots were being taken of them while they combed their hair. It is impossible to discover if what he thought was true. The evidence of Evans's pictures suggests that something different happened, that the families felt pleasure, embarrassment, pride—the emotions character-istic of many trusting family photograph sessions. But whatever the case, Agee thought the farmers naïve. He faced the responsibilities of Evans's aggression, faced, that is, the knowledge that the camera, no matter who held it, was a weapon.

What must have redeemed the situation for Agee was Evans's man-ner and pose and the dignity and complicity he allowed the families in making their own portraits. He talked to the various people, he told them his intent, he asked their permission, he let them arrange them-selves comfortably, he took the pictures only when they were ready. And he himself stood openly facing them.

This procedure was incredibly important to Agee; for from these scenes of a photographic artist and his subjects, he was able to construct a new set of priorities for art that could justify the damage he felt to be the inevitable result of prying. For Evans, standing openly with his camera, provided a visible tableau of a more hidden dynamic; he showed Agee as graphically as possible that art could be built on face-to-face encounters, that in these matters, personal influence, however profound or dangerous, could be reciprocal.

Agee extended this insight into a systematic prescription for him-self. He understood that one need not pretend to be the camera itself—a mechanical recording device—to do justice to hallowed experience. As an artist one could retain a fully human identity as long as one understood the making of art to be an act of raw empathy. Self-expo-sure justified the exposure of others.

With this guiding principle, Agee began to explore and record the nature of the mutual regard, the looking back and forth, that charac-terized his stay with the three families: First he noticed and dramatized himself:

> I know [George Gudger] only so far as I know him, and only in those terms in which I know him; and all of that depends as fully on who I am as on who he is ... for that reason and for others, I would do just as badly to simplify or eliminate myself from this picture as to simplify or invent character, places or atmospheres (239-40).

But when he looked, he himself was perceived. There were always, as he came to realize, two fields of vision. In imagination, a house could stare at him: "We stand first facing it ... and it stands before us, facing us, squarely in front of us, silent and undefended in the sun" (137). In fact, the children did stare: "And already though as yet I scarcely realize it, we have begun this looking-at-each-other of which I am later to become so conscious" (368). And finally, looking at the unforgiving face of Mrs. Ricketts at her door, he recognized that her eyes "stayed as a torn wound and sickness at the center of my chest" and that "we shall have to return, even in the face of causing further pain, until that mutual wounding shall have been won and healed" (378). The key to his persistence in an endeavor he found morally problematic was the idea of mutual wounding, for he believed that openness and susceptibility could allay fear and suspicion "through ultimate trust."

The importance of this tenet, and also its tentativeness, is suggested in the text by Agee's silent, recurrent begging. He entreated everyone at one time or another to recognize that he intended no harm and so must not be held culpable. It must be noted that these entreaties were for forgiveness as well as understanding, for Agee knew that both he and Evans had violated their own best intentions—Evans by stealing images of people unaware, just as Margaret Bourke-White had done, and Agee by similar stealth, catching glimpses of things not intended for his eyes—he was an endless snooper when people were out.

These slip-ups do not prove the rule by exception. Rather they expose the conceit of Agee's whole formulation of equality and the possibility of mutual wounding. Agee and Evans always had the upper hand in these personal negotiations and even their mildness and caring were composed attitudes, postures assumed out of respect and delicacy but assumed nonetheless to avoid the violence or excesses that were, in other less sensitive circumstances, qualities of their characters.

It could be said that Evans's posture with his camera had given Agee an ideal which, in the way of ideals, told him how to approach without giving him the qualities needed for success. Acting equal was the closest he could come to equality itself, although the respect that motivated his guarded conduct was real, as was the pain of knowing that he did, in fact, possess superior opportunities.

If Agee eventually saw the limits of Evans's photography as a model for action, so he realized also its circumscribed ability to render experience. In arresting movement, in presenting the simultaneous complexities of any given moment, the camera was masterful. Nevertheless Agee ultimately faced the same dilemma about fragmentation that had plagued Dos Passos; in his opinion photographs could not present backgrounds and contexts that were adequate. The print was always limited by a frame, whereas Agee came to feel that no frame or limit was possible in presenting and explaining his three tenant families' lives. His aggressive use of colons to punctuate the text in place of periods can be interpreted as one of Agee's answers to the framed photograph and the self-contained explanation. In his mind, one thing led into the next, each explanation inadequate to the immensity of the lives of those with whom he had lived for that one summer. If one section of his text explained "Money" and another "Shelter" and another "Work," each section must be seen as leading into the next; inseparable, inadequate to the whole, and finally, unfinished.

Agee's strongest gesture of humility in *Famous Men* was to show the poverty of language when held against the lives that preceded and inspired his text. Nonetheless, facing that poverty, he tried to write decently. "I am interested in the actual and in telling of it," he said, "and so would wish to make clear that nothing here is invented" (244).

In lessons of "real importance," Evans, with his intense regard for the "object" in front of the lens, had guided him by providing a formal artistic vocabulary of respect. That is, he had shown Agee the political and ethical dimensions of non-selectivity, of "art" that was important because of its subject's importance, not because of formal qualities imposed by the artist's shaping sensibility. Also, Evans had demonstrated a mode of reciprocal influence that could mitigate the guilt of using the "materials" gained in observing an undefended group of people. If Margaret Bourke-White had unwittingly acted with the reductionist tendencies and condescension of the middle class dealing with its inferiors, Evans had at least pointed to the possibility of art as a mode of mutual confidence.

That these lessons were only partially usable probably did not surprise Agee. He seemed always to anticipate his failure and encompass that failure in a larger acceptance. What he was trying to do in recording "the cruel radiance of what is" was ultimately beyond anybody's ability to achieve. "I have not managed to give their truth in words," he admitted, but then went on to exonerate himself and Evans by telling the story of a fox crying out to its mate in the deep Alabama night. To him the unseen animal represented the mystery of everything that lay outside the range of lens or language: it had a beauty and sadness that he could only suggest. As the sound punctuated the distant countryside, Agee described it as plainly as possible, and with all the careful attention he had directed toward the Gudger, the Ricketts, and the Woods families. But he knew that he could not understand what he heard.

Walker Evans. Sunday Singing, Frank Tengle's Family. Hale County, Alabama, 1936 *(Library of Congress)*

Bill Baker. "When Mr. Reese gets out his fiddle, he and Ruth strike up a tune. Family musicales are a favorite evening's amusement at the Reese home." From Mildred Gwin Barnwell, *Faces We See* (Gastonia, N.C.: Southern Combed Yarn Spinners Association, 1939).

Margaret Bourke-White.
"Sometimes I tell my
husband we couldn't be
worse off if we tried."
Blythe, Georgia, 1937 *(Life
magazine © Time Inc.)*

Walker Evans.
Floyd Burroughs and
Tengle Children.
Hale County, Alabama,
1936 *(Library of Congress)*

Dorothea Lange. Drought refugees from Oklahoma camping by the roadside. A family of seven, they hope to work in the cotton fields. The official at the border (California-Arizona) inspection service said that on that day, August 17, 1936, twenty-three carloads and truckloads of migrant families out of the drought counties of Oklahoma and Arkansas had passed through that station by three o'clock that afternoon. Blythe, California. *(Library of Congress)*

6

John Steinbeck
and Dorothea Lange
The Surveillance of Dissent

Nothing is more typical of so-called 'concerned photography' than the brutality of its compassion.
—Max Kozloff

I

IF Walker Evans used his talent quietly, belligerently, and in
implicit rebuke to the Farm Security Administration, Dorothea
Lange worked comfortably and in harmony with its goals. Years
later, remembering Evans's abrasive ways, she appreciated the solitary
and distinctive cast of his vision, his "bitter edge,"[1] but she knew her
own character and inclinations to be of a different tenor. Where Evans
had cast a cold eye on Roy Stryker's catalogues and files, preferring
history to take an unofficial form, Lange had eagerly espoused utility.
For her, photography was properly embedded in a larger social discourse; it extended recognition to the dispossessed, to those without
voice, to those whose absence from political power was emblematized
by their absence from images of their condition. To photograph was to
entertain the claims to attention of those whose knowledge of the
underside of American life was otherwise disqualified, diminished, or
suppressed. Where Evans separated himself from his sponsoring agency
to speak more freely in behalf of such people, Lange moved increasingly closer to her official identity.

Her story seems, consequently, to be a narrative that shows the full, self-conscious, and self-satisfactory yoking of the camera, social science investigation, and government policy. In the course of a decade, she moved from being a portrait photographer, to documenting the migrant laborers of rural California in the 1930s, to recording the Japanese interned in U.S. government camps during World War II. She began her career as a photographer of a private sensibility and ended the first, active phase of that career as part of a miliary surveillance team under the Office of War Information.

In looking back, Lange was satisfied with the logic of her own achievement. She saw a pattern; she rested in it. Those who knew her confirmed her perspective: she was, to them, extraordinarily sensitive and astute in working with her subjects; she sensed what kinds of photographs could work to their mutual advantage—a lovely person, a delicate touch, a sense of humility in the face of talent and of history. She had simply been where things had happened; she had recorded them. Her colleague, Carl Mydans, had a similar sense of the intersection of the individual and history. At the end of his memoir, *More than Meets the Eye,* he remarked, "All of us live in history, whether we are aware of it or not, and die in drama. The sense of history and of drama comes to a man not because of who he is or what he does but flickeringly, as he is caught up in events."[2] But then he told the story of an ancient, American woman who had lived through the revolution in China "unaware of her own drama."

There is a way in which this observation can be used with regard to Dorothea Lange as well, for she lived as a part of history, not simply as a recorder of it; and in the extraordinary span of her achievement, we can see the changing relation between an individual photographer and her official identity, the growing tension between a private sensibility and the needs and intentions of a sponsoring agency. Lange's work for the Farm Security Administration promoted a certain type of public visibility, and though she initially endorsed that activity, it led, logically, into another kind of supervisory observation which she later found to be much more problematic. Photographing the Japanese for the Office of War Information was not the same as photographing migrant laborers for the Farm Security Administration. Neither of these projects was analogous to photographing wealthy patrons in a studio. Each assignment had seemed to grow logically from the previous one; but in the progression, fundamental values were changed. She, as well as her subjects, was "caught up in events." To observe the sequence of Lange's career is to see more clearly the ways that a photographer's private vision of a benevolently transparent society can be transposed by a sponsoring agency into a mode of social control; it is to understand how visibility can become a mode of domination, how surveillance can grow into "malveillance," how the desire to help can increase mistrust. Her life's work brings into focus many of the themes latent, but not fully developed in the activities and achievements of her predecessors.

John Steinbeck never worked with Dorothea Lange, but he followed in her footsteps—in his fieldwork in California, in his social attitudes, and in the books which most clearly established his reputation as a fiction writer and social commentator, *In Dubious Battle* and *The Grapes of Wrath*. Both artists were greeted with great acclaim and both were occasionally scorned. Their work was respected, but it could also, at times, arouse vitriol. The resistance to their efforts is of equal importance to their eventual fame. As Lange said to Richard Doud in 1964, "[Y]ou go into a room and you know where you're welcome. . . . Sometimes in a hostile situation you stick around, because hostility is important."[3] The welcome, the hostility, and the private and official reasons for them are the subject of this chapter.

II

The general shape of Dorothea Lange's career as a photographer is well documented.[4] She came to California as a young woman, established herself as a portrait photographer, and, swayed by a sense of her own historical moment, left the security of that profession to document the life of the San Francisco streets, and, eventually, the life of the open road for Americans displaced by the Depression. At first she was motivated by something close to instinct: a day had come when the breadlines outside her studio seemed more compelling than did the portraits she customarily made; the faces that registered hunger seemed more important than did the composed expressions of those whose money defended them from want. In this early, unschooled impulse, we can see Lange's attraction to the undefended, for they seemed, in their vulnerability, to manifest the workings of a larger national pattern. What John Berger has said of Paul Strand might also be said of Lange: "The photographic moment . . . is a biographical or historic moment, whose duration is ideally measured not by seconds but by its relation to a lifetime. . . . Where he [she] has chosen to place it is not where something is about to happen, but where a number of happenings will be related."[5] Later, Lange, like Paul Schuster Taylor, would speak of the weather, banking practices, and farming procedures as "forces," equating natural acts and policy decisions; but from the first she was fascinated by those whose very persons seemed to register something larger. Her photographs were intended to show the effects of power; she intended them to be seen as part of a dialogue between what could be imaged in the immediate world, and the unseen origin of that image.

Given these early inclinations, it does not seem fortuitous or surprising that she should meet Paul Taylor, then a professor of economics at Berkeley, a specialist in farm labor practices, and a man who was increasingly concerned with problems specific to contemporary migrant workers in California. In fact, it was Taylor who noticed her photographs in an exhibition organized by Willard van Dyke in Oakland (1934), and sought permission to illustrate an essay in *Survey Graphic* with her work on the general strike in San Francisco.[6] He recognized an immediate congruity of interests and understood her photographs to offer a kind of evidence that had a compelling rhetorical effect. Of their first personal meeting and first collective project, a trip to a U.X.A. sawmill, Taylor said, "My own interest was in encouraging the photographing of a social phenomenon; the interest of the others [Imogen Cunningham, Mary Jeanette Edwards, and Preston Holder were there as well] was in finding opportunity to photograph people in social situations without fear that their motives would be misunderstood and their approaches resisted."[7]

It is not clear if Lange would have agreed with Taylor's assessment of the project. Eventually she came to see the existence of a sponsoring agency as an important personal resource in her field work, but in 1935, on first meeting Taylor, she was still an artist with private interests; she did not yet understand the social contexts of her work nor could she identify her own deepest motives. Her impulses seemed more nebulous and, at this point, distinct from those of the social scientists she had just met. It is true that the strikers and men on the streets had frightened her: "At that time," she remembered, "I was afraid of what was behind me—not in front of me. . . . Really it's my camera I was so afraid about. I thought someone might grab it from the back and take it away or hit me, from the back."[8] She did not, however, clearly identify this anxiety with the inherent nature of her private photographic enterprise and consequently did not feel the need of an official status to shelter her from the suspicion of her subjects. Her own memories of these early trips are memories of entering a new world. Though Taylor might have understood the sociological value of her photographs, she did not consider herself an investigator. In fact, she was surprised by investigative attitudes and methods. Taylor was then Field Director of the Division of Rural Rehabilitation (California State Emergency Relief Administration), and when he and one of his students, Tom Vasey, took

Lange on a trip to record the pea harvest in Nipomo (February 1935), she said:

> I remember ... the first day we were out in a car—we'd stopped at a gas station—Paul asked the fellow who put the gas in the car some question about the country around—as we drove off Tom said, "He was a good informant." ... I thought, "What language! What kind of people are these? 'He was a good informant.'" That really surprised me. I knew then that I was with people who were in a different world than mine.[9]

But it became her world, both personally and professionally, for in December 1935, she married Taylor and continued the collaboration that focused national attention on the situation of the migrant workers and established their reputation as a documentary team. If she began this work instinctively, she grew into an understanding of it which accorded with Taylor's sense of his own intellectual mission. For both of them, information functioned as Rousseau would have wanted it to function—to make society transparent, to take power out of the unseen but seeing eyes of an elite and place it firmly in the hands of the collective social body. As Foucault has said, "It was the dream that each individual, whatever position he occupied, might be able to see the whole of society, that men's hearts should communicate, their vision be unobstructed by obstacles, and that opinion of all reign over each."[10] For Taylor, "the opinion of all" was represented in the United States Congress, and it was in address to the state and federal governments that he wrote the reports that first incorporated Lange's photographs within them.

Lange and Taylor established a close working relationship very quickly. The first trip to Nipomo for the pea harvest occurred in February 1935; the two reports that later became known as the Taylor-Lange Reports were submitted on 15 March and 17 April of the same year. Both were written for the California State Emergency Relief Administration (S.E.R.A.). The first spoke of the peculiarities of California's agricultural economy—about its high degree of capitalization, its growing concentration of land ownership, and its inordinately high dependence on migrant farm labor. Its function, however, was to explain only insofar as understanding would promote action. Taylor saw a radically polarized structural situation, where the centralization of land ownership jeopardized labor negotiations with individuals, leaving single workers with no effective voice. He was struck by the silence of the workers and came to see their strikes as the only rhetorical mode open to them. "Against these conditions there is no vocal protest pos-

sible from the laborers themselves, but fifty-odd strikes have torn California agriculture since December 1932."[11] His first report was
intended to achieve through language what the strikers sought to
achieve through demonstration, to substitute advocacy for protest. Since
he located the cause of their unrest in ludicrously inadequate housing,
he located its cure in improved living conditions. Quoting the U.S.
Special Commission on Agricultural Disturbances in Imperial Valley
(11 February 1934), he announced:

> We found filth, squalor, an entire absence of sanitation, and a crowd-
> ing of human beings into totally inadequate tents or crude structures
> built of boards, weeds, and anything that was found at hand to give
> a pitiful semblance of a home at its worst.... In this environment
> there is bred a social sullenness that is to be deplored.

He added, "The attached photographs taken in early 1935 by the Division of Rural Rehabilitation document this condition which words are
inadequate to describe."[12] Taylor had assembled this material, along
with cost analyses, as part of a proposal for government subsidized
camps for migrant workers, a project funded through its experimental
stages but then dropped.

If Lange was alienated when she began her work with Taylor, she
swiftly overcame her estrangement. Taylor had literally embedded her
work in the context of his report, but she came quickly to identify
with that context and to assume attitudes and photographic goals consonant with Taylor's economic analyses and recommendations. His second report was an attempt to show that the Oklahomans arriving in
California were part of a major demographic shift in America, itself
the result of the newly mechanized farming practices used in the Midwest. Years later, as her own explanation of the midwestern exodus,
Lange reiterated Taylor's explanation of the migrants' appearance.
After a while, it becomes difficult to distinguish their thinking. Lange,
for example, claimed credit for having first noticed the Oklahomans'
arrival:

> The influx into California after the dust storms of April 1934 [1935?],
> I made the first report on! The first wave of those people arrived in
> southern California on a weekend. It was as sharp and sudden as that
> when I was there . . . that Sunday in April of 1935 was a Sunday that
> I well remember because no one noticed what was happening, no one
> recognized it. A month later they were trying to close the border.[13]

She knew the travelers to be in trouble. Her work's usefulness, like
Taylor's, became its *raison d'être*. She used the camera as an extended eye

of power, uniting, metaphorically, two parts of the social body that
ignorance or neglect had separated.

Walker Evans and Margaret Bourke-White might have identified
their own documentary goals in a similar fashion, but it is important
to notice that Lange was the only one of them to have actively
embraced a government agency as the appropriate arena for work.
Bourke-White had worked privately in the rural South; Evans apparently had obscured his role in the Farm Security Administration while
in the field, preferring to seem anonymous or unattached to any particular policy. Only Lange was comfortable, and, in fact, eager to have
an official sponsor and to make it known that she had one. She was a
pragmatist. She did not attribute force to opinion alone. Congress, not
an amorphous public, was her ultimate witness. In her own estimation,
the government's interest justified her presence among the migrant
workers; it clarified her relation to them and forestalled the problems
of relationship that had hounded her predecessors. She experienced
none of their anxieties about ulterior motivation. "When you photograph," she said, "all you're doing is a job ... and you have every
right—*must* have every right in your mind—to make that photograph."[14] Even when asked quite pointedly if she had ever experienced
hostility or resentment in the field, she denied it, preferring to remember the migrants' reception of her as open and grateful. How often, she
asked, does anyone get "the full attention of the person who's photographing you[?] It's rare, you don't get it very often. Who pays attention to you, really, a 100 percent?"[15] That it was the government's attention, as represented by her professional status, seems to have offered no
complication. She believed in what she was doing and found confirmation of her beliefs in her easy reception in the field.

> They know that you are telling the truth. Not that you could ever
> promise them anything, but at the time it very often meant a lot that
> the government in Washington was aware enough even to send you
> out. And there were times along then when the photographs were
> used in Congress so that you could truthfully say that there were
> some channels whereby it could be told.... People are very, very
> trusting.[16]

In fact, one might observe that Lange did much to engender a belief
in the benevolent eye of power; for in working, she enacted her version
of a democratic society composed of an equal but diversified people.
The migrant's role was to pick fruit; the photographer's role was to
communicate, to convey information, to act as a conduit. The Con-
gressman's role was to allocate collective funds. Because governance

was perceived to be by and for equals, to give visual information to Congress was simply to establish an informed citizenry. Never did she think or act in hierarchical terms; never did she understand her work to be part of a scheme for tactical control through surveillance. If anything, she and Taylor considered themselves, ironically, to be engaged in counter-surveillance, since those who wanted to observe the migrants in the interests of domination and suppression were presumed to be in the Growers' Association, not in the U.S. Congress. As they saw it, farm owners, concerned to maintain low morale, low wages, and the isolated powerlessness of the workers, were anxious to have information about them, but anxious, also, to suppress the dissemination of that information. They wanted to keep the workers as the objects of discourse, but not to let them participate in or originate discourse.

Lange feared that power was being taken out of the hands of what she considered legitimate authority, taken away from ordinary people and vested in special interest groups. In her eyes, vigilantes—those who watched as tools of private rather than of collective decision making— were those she guarded against by offering another version of social reality with her camera. She worked to expand the world of social discussion, to bring into discourse a suppressed version of the truth about the migrants' living situation so that, at the least, conflicting truth claims could be weighed. "Whose cognition of events in the farm labor situation in California would predominate?" was the central issue she addressed by taking the camera into the fields among disheartened and disenfranchised people. "I feel myself more like a cipher, a person that can be used for a lot of things, and I like that," she said.[17] Foucault would have insisted that she entertained the claims to attention of an otherwise disqualified knowledge.[18]

What Lange inscribed on film, Taylor inscribed in language. The two modes of knowing became, at times, inseparable. Taylor remembered their way of working together in the camps along the highways:

> Out of the corner of my eye, I would see that she had got out her camera, so I would just keep the talk going. I had their attention . . . keeping out of range generally wasn't a problem. She worked pretty close to them for the most part. If I thought I was interfering, I would just sidle out of her way as inconspicuously as possible, talking to them all the time. My purpose was to make it a natural relationship, and take as much of their attention as I conveniently could, leaving her maximum freedom to do what she wanted.[19]

If this is accurate, it tells us that many of Lange's photographs are images of one side of a dialogue from which Taylor had just moved:

having initiated the conversation, he stepped aside, leaving the speaker in the center of the frame, alone, but engaged in reciprocal interaction.

Taylor's presence may have multiplied the modes of interaction among the farm workers and investigators, but it did not change fundamentally Lange's approach to her subjects. She was, she said, "a photographer second."[20] What people gave to her and what she gave to them were the prerequisites of any photographic session: "We found our way in, slid in on the edges. We used our hunches, we lived, and it was hard, hard living." Her memories of working are memories of making connections and of establishing a personal identity among strangers. There were no unwitting or naïve moves on the part of those photographed. They knew Lange; they knew why she was there.

> You know, so often it's just sticking around and being there, remaining there, not swooping in and swooping out in a cloud of dust; sitting down on the ground with people, letting the children look at your camera with their dirty, grimy little hands, and putting their fingers on the lens, and you let them, because you know that if you will behave in a generous manner, you're very apt to receive it. . . . And I have asked for a drink of water and taken a long time to drink it, and I have told everything about myself long before I asked any question. "What are you doing here?" they'd say. "Why with your camera? What do you want to take pictures of us for? Why don't you go down and do this, that and the other?" I've taken a long time, patiently, to explain, and as truthfully as I could.[21]

> You have to wait until certain decisions are made by the subject— What he's going to give to the camera, which is a very important decision; and the photographer—what he's going to choose to take. It is a much larger inner process than putting the camera between you and the subject.[22]

It is this combination of traits—an unassuming nature coupled with an openness fostered by the official endorsement of her task—that defines Lange's particular genius. No matter how much she may have believed it herself, it was not official identity alone which ensured the effectiveness of her documentary work. Although she attributed her success to the structure of the situation—the federal government's concern, through her, for the welfare of its citizens—the structure could function malevolently as well as with good will. She discovered this in the early 1940s when the Farm Security Administration was turned into the Office of War Information (O.W.I.), when her own work on the farm labor problem ceased and she began to use her camera in the service of the War Relocation Authority.

Her major assignment was to photograph the Japanese who were being interned in government camps for the duration of the war. She recorded them as they left their communities, traveled, and settled into the structures the government had erected for them to live in. In her memory, it was an assignment hampered by elaborate security and clearance procedures: there were some things she could not photograph; and nothing could be published independently—the O.W.I. owned all of the negatives and the Wartime Civil Control Agency impounded many images until the completion of the war. As Karin Ohrn observed, "[A]lthough the army wanted a record, it did not want a public record." Lange herself soon realized the secretive nature of her own work: "None of it extended outside the watchtower."[23]

The irony of Lange's participation in the O.W.I. lies in the patterns of reiteration and reversal that were embedded in the structure of this work. Where the Oklahoma migrants were homeless because of the "forces" of nature, consolidated money, and mechanized farming practices, the Japanese were homeless because of the "forces" of government policy concerning national security. Neither group was in transit voluntarily. Each bore the impress of some larger decision-making process. Each was housed by the federal government in camps. But where the farm camps at Arvin and Maryville were places in which First Amendment rights were observed, where walls meant privacy and security against the surveillance of the Growers' Association, the Japanese camps were experienced simply as prisons. Lange's camera became the visible extension of a supervisory authority that needed to monitor the activities of "aliens" without letting their activities be more generally known. Though the O.W.I. had grown out of the F.S.A., its metamorphosis was not simply a logical expansion but an ideological shift from Rousseau to Bentham, from the ideal of a transparent society governed by the opinions of an enlightened citizenry to the reality of a disciplinary state governed by an unseen and secretive central authority. This shift was occasioned by the war, but it meant that Lange's role had changed fundamentally. From being a conduit of information in an open situation she moved to being a means of supervision and control. With regard to agricultural workers, the evidence of photographs had worked in the service of both subject and audience; with regard to the Japanese, photography could have no beneficial effects. It related them only to the power that continued to intern them.

In fact, the potential for suppression had always existed in the structure of social science inquiry itself. Although Lange always photographed with a strategic purpose, she worked benevolently, and since her own attitudes seemed to her initially to reflect those of the Congress which had created her sponsoring agencies, she did not understand herself to be engaged in anything questionable. Her own unassailable motives buffered her from seeing the dubious motives of others. This perspective changed when she worked at Manzanar, the Japanese relocation camp. It is to Lange's credit that she considered this internment a misuse of legitimate authority. It is one of the puzzles of her career that she persisted in her work despite the perceived illegitimacy of the agency that sponsored her.

III

Dorothea Lange's photography was initially embedded in Paul Taylor's economic and social analysis; it eventually became embedded in John Steinbeck's novels. His halting reliance on her work can show us a great deal about the catalytic effect of one art form on another, as well as the larger transformations of politics into culture.

If Dos Passos and Agee had found creative influences in the working methods of photographers with whom they actually lived, if the dedication of their lives had preceded the transformation of their art, Steinbeck experienced no such immediate personal relationship. His novels grew, in part, out of the photograph's ability to promote a surrogate sense of experience, and though he followed in Lange's footsteps, traveling some of the routes that she had taken on her photographic tours, he never went out in the field with her. They seem to have known of each other for years without finding occasion to interact personally.

This circumstance is critically important, for Steinbeck used Lange's photographs in researching his novel, *The Grapes of Wrath*, and he assumed, with almost uncomfortable closeness, the ideological positions of the Taylor-Lange Reports on farm labor conditions; but he failed, almost entirely, to understand or to emulate the working methods that Lange assumed as a consequence of her beliefs. One could say that Steinbeck mimicked a formal aesthetic posture without penetrating the rationale for its genesis. He, as did Lange, had many occasions to reflect on the codification of power relationships through sight; but though he could see the belligerence engendered by the act of surveillance, he seemed to be oblivious to the effects of his own creative tactics.

The book which has occasioned the most interest in Steinbeck's relation to Lange is *The Grapes of Wrath*. Published in April 1939, it marked the culmination of years of concern with migrant labor unrest in the central valleys of California. Where did the book come from?

To what extent was it fabricated? What kind of experience or direct observation lay behind its descriptions of families exiled by the Oklahoma Dust Bowl, harassed by California growers, sheltered by federal housing, and pitted, finally, against flood and natural disaster? These questions of origin, still imperfectly sorted out, are naïve in the sense that they assume realism to have an origin in experience. They also skirt the most interesting problems of observation posed by the text, although they do provide a useful place to begin addressing other more complicated issues.

Interest in the origins of *The Grapes of Wrath* began soon after it was published, but one of the most intriguing speculations was generated in 1974 when D. G. Kehl made a case for considering that Steinbeck had based his novel on a "string of pictures" he had seen in two documentary books. One, *An American Exodus,* was by Lange and Taylor and the other, *Land of the Free,* was by Archibald MacLeish, who used the F.S.A. photography files to supplement his poetic commentary. It was an interesting hunch, one that I now think was incorrect, but it suggested a connection between Steinbeck and the F.S.A. The hypothesis was based on a similarity between the pictorial quality of Steinbeck's prose, and photographs of analogous situations: Lange had toured Oklahoma and the Southwest in 1936, making images of abandoned farm houses in the Texas panhandle and of people in transit. Her photographs of Sallisaw, the hometown of Steinbeck's fictional Joad family, show men squatting in town in consultation with each other and, after having made the decision to leave, at the wheels of old cars loaded with mattresses, coffee pots, and large families. Kehl followed his description of each phase of Steinbeck's book—the "blown out" Oklahoma farms, the journey to California, the wretched Hoovervilles, the government camps—with the admonition: "[T]his should be examined in conjunction with Lange's photographs in *Land of the Free* and *An American Exodus.*"[24]

The hypothesis falters on questions of chronology and motivation. *Land of the Free* was published in 1938, *An American Exodus* in 1939. Steinbeck was at work on a second draft of his novel by late 1937. If he based his book on pictures seen in 1938 and 1939, what did he write in the preceding years? Did he ever actually read *Land of the Free?* The evidence is circumstantial at best, for MacLeish, when questioned, said only that Steinbeck "may very well have seen [it]."[25] With regard to *An American Exodus,* the evidence is certain, for Lange had asked Steinbeck first to introduce her work to his publisher, Pascal Covici, and later, when Reynal and Hitchcock agreed to publish it, to write a preface to it.[26] But these speculations become idle when the chronology of Steinbeck's involvement with the migrant labor problem is uncovered, for he knew about Lange's and Taylor's efforts long before this, and in fact it was knowledge of their fieldwork that spurred him to make firsthand investigations of his own. He had no reason to read either *Land of the Free* or *An American Exodus,* for, by 1939, he had already followed a step behind Taylor and Lange, using their reports, going where they had gone, producing newspaper articles and tracts about the migrants, which can be described as popular interpretations of the policy papers Taylor had written for the S.E.R.A. One of Steinbeck's essays, a collection of newspaper articles called *Their Blood Is Strong* (1938), was even illustrated, as Taylor's reports had been, with Lange's photographs.

This "echoing" began in the summer of 1936, when Steinbeck went to the Division of Information Offices of the Relocation Authority of California to gather statistics for some essays on the farm labor situation. Already known for the empathy shown to laborers as the author of *In Dubious Battle,* he was encouraged to visit the federal government's first residential camps for migrants. He made a tour in the last two weeks of August with Eric Thomsen, the regional director in charge of management.[27] The first camp, near Maryville, had been set up by Tom Collins. The second, in Arvin, was currently managed by him. A year after its construction, Steinbeck walked into Weedpatch, one of the camps which Taylor and Lange had worked so strenuously to establish in 1935.

This visit marks the beginning of Steinbeck's "trailing" of Lange. In early October 1936, he went to the strike areas of Salinas and to Bakersfield, the town which marked the point of entry into California for most of the western migration. From 5 to 11 October 1936, the *San Francisco News* ran a series of articles by Steinbeck called "The Harvest Gypsies." These are the pieces that were later gathered together and published by the Lubin Society as *Their Blood Is Strong.* The pamphlet was illustrated with Lange's photographs, but, more importantly, it was informed, almost to the issue, by the statistics and attitudes of Taylor's and Lange's 15 March and 15 April S.E.R.A. reports: Taylor had seen the Oklahoma migration as part of a great demographic shift occasioned, on one side, by the mechanization of farming in the Midwest and, on the other, by the peculiar structure of California's agriculture. Steinbeck pointed out "the unique nature of California agriculture" where growers required migratory labor at the same time that they denigrated the people who did it. Taylor spoke of the "high degree of farm capitalization" and the "concentration of land ownership." Steinbeck called the same phenomenon "corporation farming." Taylor quoted the U.S. Special Commission on Agricultural Disturbances, pointing out that people lived in "inadequate tents or crude structures built of boards, weeds and anything that was found at hand" and that "social sullenness" was bred by deplorable living conditions. Steinbeck described the ramshackle squatters' camps and observed that "spirit has turned to sullen anger before it dies." He then passed on Taylor's statistics about the cost of establishing the labor camps at Arvin and Maryville, the sanitary provisions needed, and the benefits in terms of "human dignity" to be derived from investing more heavily in such projects.

His only original contribution to telling the story of migratory labor was the addition of case studies to illustrate Taylor's more abstract points. He vivified the migrants by telling about the experiences of several families who were ill-housed: "With this [child's] death there came a change of mind in this family. The father and mother now feel that paralyzed dullness with which the mind protects itself against too much sorrow and too much pain."[28] Where Taylor resorted to photographs for rhetorical effect, Steinbeck turned to anecdote, alternating comments about a collective situation with description of the people forced to live out individual fates within the larger pattern of national disaster.

In this early set of essays we can see the seeds of Steinbeck's commitments in *The Grapes of Wrath*. Though he did not know it then, the essential features of his novel were in place by 1936: he had a set of antagonisms and endorsements. The novel repeats still another time Taylor's explanations of the causes of agricultural distress; it reiterates the villainy of corporate farming and the salvation to be found in government-sponsored migrant camps. It embellishes on Taylor's plea for Congress to notice that these particular itinerant laborers are decent Americans whose heritage entitles them to consideration not previously given to Mexican or Japanese fruit pickers. Like Taylor, Steinbeck lionized the common American, seeing the migratory worker or the boy in the air force as a variation of the pioneer, "possessing and living in close touch with the land ... descendants of men who crossed into the Middle West, who won their lands by fighting."[29] And the novel uses the structural principle discovered in *Their Blood Is Strong*: it builds an epic out of alternative focuses. Steinbeck presents the whole phenomenon of "exodus" and then, in intercalary chapters, turns to its effects on the lives of the Joads, a fictionalized family from Sallisaw, Oklahoma. In 1938, in the midst of writing, he spoke of his technique as the "little picture" and the "large picture": "Mustn't think of its largeness but only of the little picture while I am working. Leave the large picture for the planning time."[30]

But in 1936, these thoughts were not uppermost in his mind. He may have been anticipating his novel, but for the time being, he wanted simply to have some effect on public opinion, and he was afraid that his newspaper reports would be one more futile gesture of goodwill. The migrants, he realized, were suspicious: "Helpful strangers," he noticed, "are not well received in this [squatters'] camp."

> The local sheriff makes a raid now and then for a wanted man, and if there is labor trouble, the vigilantes may burn the poor houses. Social workers have taken case histories. They are filed and open for inspection. These families have been questioned over and over about their origins, number of children living and dead.
>
> The information is taken down and filed. That is that. It has been done so often and so little has come of it.[31]

His observation was supported by a letter from the Camp Committee at one of the federal camps:

> Dr. Mr. Steinbeck,
> We saw your letter to the Editor of the San Francisco news exPlaining Why you used the word Gypsies.
> We all understand just Why you found it important to use that word, we know their are lots of People who know from seeing just how Gypsies live, but we also Know that their are more people that dont Know how farm workers live, and never would know if it had not been for your trying to exPlain and Show them.

They excused him because they sensed his blundering good will:

> We think you did a fine job for us and we thinkyou. this is a big battle which cannot be won by ourselfs. We kneed friends like you to help us get decent camp places and after that we can handle the farmers ourselfs by working together as we have learned to do at this camp.
> > Your friends
> > Camp Central Committee[32]

Troubled by this letter, perceiving that his forgiveness was partial and that he himself was considered a "helpful stranger," he left for a trip to Europe in the summer of 1937.

When he returned to the United States in September, he came with renewed interest in his novel. Before traveling to the West Coast, he went to Washington, D.C. to look through the full files of the Farm Security Administration. His use of these photographs was a curious one. On one hand, the pictures served as a surrogate for experiences he had never had: though he later led his friends to believe that he had

gone to Oklahoma on a research trip, he actually never left California
and could have constructed the Joads' fictional Oklahoma life only
from secondhand sources like the F.S.A. file. On the other hand, the
photographs spoke to him of their own limitations; they suggested the
inadequacy of surrogate experience and the need to face, for himself,
what Lange had faced with her camera. If Steinbeck had trailed Lange
and Taylor ideologically, assuming their perspectives and using their
statistics, he trailed them literally in asking the F.S.A. to send him into
the field. Charles Benson quotes the letter that finally clarified the
arrangement Steinbeck made with the F.S.A.: he was not hired by the
agency, but, as Benson has pointed out, it subsidized his work indirectly. Dr. Will Alexander, deputy administrator of the F.S.A., made
the arrangement, and C. B. Baldwin, who was then Alexander's deputy, remembers what happened:

> Steinbeck told us he wanted to write a novel about migrant workers,
> but, as he expressed it, he needed the experience of a migrant worker
> if it was to be a realistic story of how they actually lived. He asked
> if we could give him this opportunity and said he would need the
> help of migrants or at least someone with whom he could work as a
> migrant. We were impressed, and because of the intense opposition
> of the Farm Bureau and the Associated Farmers in California this
> entire program was threatened, we were anxious to cooperate. I called
> Jonathan Garst and he suggested assigning Tom Collins to work with
> Steinbeck. Tom responded enthusiastically. As a result, Tom Collins
> and Steinbeck became a two-man team, both working in the Imperial
> and San Joaquin valleys as migrant workers for some months.[33]

From this point, Steinbeck's reiteration of Lange's fieldwork became the
mark of his own misunderstanding: where Lange used the F.S.A.'s spon-
sorship as a supportive presence, as an endorsement of her right to be
among strangers, Steinbeck used it only as the agent of disguise. Where
she announced her purpose through the purposes of the government
which sent her on assignment, Steinbeck obscured his purposes all
together. His connection with Tom Collins was unseen and unan-
nounced. In October 1937, preparing to visit Collins at the F.S.A. camp
at Gridley, Steinbeck wrote to him, saying, "Don't tell anyone I am
coming. My old feud with the ass[ociated] farmers is stirring again and
I don't want my movements traced."[34] Collins kept his vow of secrecy
and later came to regret it deeply. In retrospect he believed that his
collusion with Steinbeck had cost him his position as camp director: it
was he, not the writer, who personally faced the responses of the
migrants to Steinbeck's disguised presence among them. Like Jack Lon-

don in the East End of paupers' London, Steinbeck had used all the tactics of spycraft: he had eavesdropped behind the lines; he had deceived people; he had urged their unwitting confidences by moving among them with an altered identity. In their eyes, he was no better than were the vigilantes who infiltrated their camps, planting social unrest through disguised workers. When writing earlier about vigilante activities in *In Dubious Battle,* Steinbeck had observed, "Mostly they wear masks."[35] Wearing their own metaphorical masks, armed with a strategic purpose and clandestine techniques for achieving it, Steinbeck and Collins "worked" the migrants. From Collins's later beleaguered report of this collaboration, we know that their vigilance was experienced by the migrants as a kind of violence; their usury was discovered and resented. Nothing in this play of author-as-worker could have been further from Lange's openhanded way of working in the field. Nothing could have been more distant from her officially embraced identity. Though Steinbeck had sought to recodify a relation of force, aligning art with the politically oppressed, using fiction as the displaced voice of the dispossessed, he ended by reduplicating the power structure he thought he was criticizing. Superficially Steinbeck's methods and ideological commitments were similar to Lange's—they both considered that they used their talents in the service of an otherwise voiceless people—but in the openness of one and the concealment of the other lay a world of difference. Steinbeck's approach to his subjects was especially curious for someone who had understood the mutual mistrust of workers and growers, and who had already written about the surveillance and sabotage used against the apple pickers in the Torgas Valley. *In Dubious Battle,* written in 1934, years before his involvement with Lange and the F.S.A., was, in certain respects, a survey of the kinds of watchfulness that buttress discipline in labor relations. "Did you ever work in a place where they talked about loyalty to the firm, and loyalty meant spying on the people around you?" (9), the central character asks in the opening of the book. For Steinbeck the issue of unseen authority disciplining through surveillance was openly considered. It was, ironically, one of the problems which underlay his own work and finally sabotaged his credibility in the eyes of those he sought to help.

IV

In Dubious Battle is the story of a fruit pickers' strike that is orchestrated by several inexperienced labor organizers. Even Steinbeck's handling of the subject was political, for, like Stephen Crane in his New York City sketches, he tried to render the action from the perspective of men who were in travail; he tried to articulate the thoughts and experiences of those who did not ordinarily speak but who had to demonstrate their sentiments. If Taylor had recognized the strikes of the central valleys as the itinerant workers' only rhetorical mode, their only eloquence, Steinbeck recognized the need to articulate the strike. "What promoted collective action?" he asked. "What controlled it?" "What did it feel like to be in the midst of it?"

To some extent, then, Steinbeck's motives in writing about migrant workers were analogous to Lange's motives in photographing them. To notice these people, to consider their viewpoint worthy of attention, was in itself to expand the realm of social discourse.[36] But there is one essential distinction between their work: where Lange rested with adding truth claims to the discussion of social issues, saying, in effect, "We must also consider this," Steinbeck went on to challenge existing truth, to suggest that new claims replace older, false ones. The "resurrection of a suppressed knowledge," the writing of a book that approximated the strikers' view of the world, called into question the validity of a more generally accepted version of the farm labor situation, a version "written" by the Growers' Association, by "a few men [who] control everything, land, courts, banks" (152).

The "dubious battle" of Steinbeck's title might be seen, then, as the struggle of conflicting claims of knowledge. "Who will tell the story of the migrant workers' rebellion?" is the question the text both addresses internally and participates in by its very existence. As Steinbeck realized, events are seen in the light of that which is itself unseen. An author is like the vigilante in the book who stops two labor organizers: as they ride through the darkness in their car, Jim, who philosophizes, who looks at the strike in the larger terms of justice, dignity, and the ends of human life, says, "Look at the stars, Mac. Millions of 'em." Mac, who is both plodding and driven, who always uses events pragmatically, replies, "You look at the road" (103). Finally neither view prevails because it is the two of them who are looked at: "Without warning a blinding light cut out through the darkness and fell on the men's faces. . . . A voice behind the light called, 'These are the guys.' . . . Because of the light beam, the man on the door was almost invisible. He said, 'We want you two out of Torgas Valley by daylight tomorrow, get it?'" It was a situation that Steinbeck was to repeat almost ritualistically in *The Grapes of Wrath*, where the deputy sheriff uses the spotlight of his car to ferret out farmers who are "trespassing" on the land they previously owned. "'Now duck,' said Muley. They dropped their heads and the spotlight swept over them and crossed and recrossed the cotton field. . . . 'Lead off,' said Joad. . . . 'I never thought I'd be hidin' out on my old man's place.'"[37]

In both instances, authority stands invisibly behind the light, using it to survey and then to control those who are caught in its beams. Light is not truth, but simply the agent of a supervisory power. Steinbeck uses light metaphorically to describe the same situation that Hawthorne described generations previously, in a political setting, and in both cases, it expresses the wish to see without being seen, to manipulate a play from behind the curtains: the stage manager, the vigilante, the writer all share analogous structural positions. But where Hawthorne described one art form the activities of which were covert in terms of another art form with observable methods, Steinbeck described his art in terms that equated art and politics, or, rather, that brought the political ramifications of art into the open: to be the authority behind the light is to control human destinies; to expose without coming into view oneself is to exercise an irrefutable power. In the case of the California vigilantes, it was the power to control the movements of men, and in extreme instances, the power to maim or kill. In the case of the writer, it was the power to define, to name, to declare a situation to be true. But, finally, the power to control and the power to define are analogous, for the legitimation crisis of Steinbeck's book is the struggle of an inarticulate people to speak in their own voice and to act in their own behalf. In short, it is about their efforts to evade the illegitimate control of others, to refuse the definitions of themselves and of life's possibilities put forward by those in authority.

The struggle of the strike organizers and growers to present things "in their own light" is thus the central issue of *In Dubious Battle*. Not only does each group want to manipulate "the mob"—either by forcing men back to work or by encouraging them to oppose collectively the wages offered them—but each needs to control how much is known by others about the situation of conflict. Within the novel, the newspapers function as the liaison between the laborers and the community, and they serve as an intratextual model for the function of the novel itself as a purveyor of truth. Within the text, they allow us to see who controls the truth claims about the situation: in one case, a blatant misrepresentation purposely suppresses the truth: "The paper carried a headline, 'Supervisors vote to feed strikers. At a public meeting last night the Board of Supervisors voted unanimously to feed the men now striking against the apple growers.' . . . London broke in, 'I don't see no reason to kick. If they want to send out ham and eggs it's O.K. by me.'

'Sure,' Mac said sarcastically, 'if they want to. This paper don't tell about the other meeting right afterwards when they repealed the vote'" (172–73).

In another later instance, the newspaper accuses the strikers of burning the farm of the man who has provided them with private shelter from the police. According to the novel, neither account tells the whole story, but each has rhetorical value: since no one will volunteer food to people who are already fed, the first article persuades its readers to cut off the strikers' food supply. And because people who were not witnesses to the fire cannot tell that it was set by vigilantes, they cannot know that this article is part of a strategy to denigrate the strikers as irrationally violent and without legitimate grievances. The article diminishes the strikers' base of public support. Steinbeck's novel is, in this respect, part of a larger rhetorical strategy which, by exposing false truth claims, by showing the deceptions of the stories within its own more encompassing narrative, endorses another version of the truth: the strikers are starving; they are peaceful victims of violence, falsehood, and deplorable working conditions. The novel polices the vigilantes in the same way that the vigilantes police the migrants. Its strategy is to discredit one official story and replace it with other information that has been repressed or discredited.

To the extent that Steinbeck's narrative accords with the perspective of those living through the events, one could say that it is a better story than the ones featured in the newspaper. But a legitimation crisis is not only a response to inadequate or harmful official definitions, but also a response to the way definitions are created. The newspapers may define the apple pickers as a vagabond and worthless bunch, but the novel eventually shows other versions of the strike to be suspect as well. The labor organizers who "author" the workers' side of the struggle are exposed as equally suspect manipulators who use the workers as part of their own plot. Though they consider themselves benevolent and believe their interests to accord with the interests of the men, Steinbeck would have us understand that their usury is commensurate with their antagonists'. Since they are inexperienced and have never led a labor action before, they are continually preoccupied with "making up" the strike. Eternal improvisers, using any material at hand, they construct their scenarios out of random violence. People are not people, but "examples." Without the authority of law and without actual weapons, human life and human emotion become the equivalent of guns: when an old man breaks his hip falling from a decayed ladder, his misfortune is used to incite anger; when Joy, a demented communist, is killed in trying to convert a bunch of scabs, his death is welcomed: "He's done the first real, useful thing in his life.... We've got to use him to step our guys up, to keep 'em together. This'll make 'em fight" (148–49).

These incidents continue until the conclusion of the text, where the reiteration of a "useful" death—this time of the young idealist, Jim—tells us that we are witnessing an obsessive recurrence, an end which is not a closure but an element of compulsion. Mac, seeing Jim dead with his face shot off, picks him up and takes his body back to camp:

> London handed the lantern up, and Mac set it carefully on the floor, beside the body, so that its light fell on the head. He stood up and faced the crowd. His hands gripped the rail. His eyes were wide and white. In front he could see the massed men, eyes shining in the lamplight. Behind the front row, the men were lumped and dark. Mac shivered. He moved his jaws to speak, and seemed to break the frozen jaws loose. His voice was high and monotonous. "This guy didn't want nothing for himself" (312–13).

This man's use of his young friend's life is as cold-blooded as any of the calculations of the Growers' Association, and this is Steinbeck's point. If one group exploits the men for profit, the other exploits them in the name of a visionary future. In each case, private motives dominate the collective will of the men: no longer within the arc of light, observed and threatened, Mac stands in darkness, turning the kerosene lantern on the ruin of a hopeful youth and on the faces of the strikers, uniting all of them in his iron vision of their proper destinies. Speaking in the mode of supremacy, speaking on his own behalf, separated by his obsession from those with whom he had sought unity, Steinbeck's organizer is a profiteer of the emotions, a trader in the commodity of life, a violator of the ethics of equality that the strike was organized ostensibly to promote.

If Steinbeck's purpose in writing *In Dubious Battle* was to discredit the Growers' Association as the teller of labor history, he did so in a curious way. For the text exposes the Associations' narrative reconstructions of events by exposing the equally heinous constructions created by their challengers. Like a Chinese box puzzle, the novel embeds one falsity within another until growers and strike organizers alike stand outside human bonds, disconnected from community. The desire to author any version of the collective life of a group to which one does not belong has turned a rite of incorporation into a rite of separation.[38] And these struggles, as Steinbeck saw more and more clearly when he later wrote *The Grapes of Wrath,* became analogies for his own predicament. At this early point in his career, he could name the violations of

reciprocity involved in the farm labor situation; he could write about
the relation of surveillance to discipline and control. But he had learned
of these things secondhand, through the confessions of two fugitive
organizers whom he had paid for their story.[39] He wanted to write *The
Grapes of Wrath* with a perspective gained from experience. It was at
this point in his life that he made contact with the F.S.A. and arranged
to follow in Dorothea Lange's footsteps. He could not yet see that his
efforts to duplicate her work for the F.S.A. would fail and that he
would instead fall into the very kinds of exploitation that had fascinated and repelled him during his own earlier writing.

V

Steinbeck's most prolonged fieldwork began when he returned from the
Washington F.S.A. office in 1937. In October he bought a used baker's
truck and visited Tom Collins of Weedpatch camp; later the two of
them traveled around the interior valleys together. Toward the end of
February 1938, he returned in the company of Horace Bristol, a *Life*
magazine photographer, to document Visalia, an area south of Fresno
where thousands of families were stranded and starved due to floods.
As he said in the introduction to Collins's unpublished novel, *Bringing
in the Sheaves,* "Later in the year Windsor [Windsor Drake, Collins's
pseudonym] and I traveled together, sat in the ditches with the migrant
workers, lived and ate with them. We heard a thousand jokes. We ate
fried dough and sow belly, worked with the sick and the hungry, lis-
tened to complaints and little triumphs."[40]
     His working and listening earned him the Nobel Prize in 1962. The
Academy recognized his work as "serious and denunciatory"; *The
Grapes of Wrath* was called an "epic chronicle" of American life. Structured according to the principle discovered in *Their Blood Is Strong,* alternating the troubles of a single family with the troubles of 200,000 similarly afflicted people, the book contained an historical perspective
rarely achieved in American fiction. The Joads took their place amid a
sweeping national cataclysm; they were, as Steinbeck said, "[A]ll of
them . . . caught in something larger than themselves" (32). Like Dorothea Lange, he was interested in the effects of power on individual life;
like her, he gave his full attention to one subject and then let the accumulation of similar images show the magnitude of displaced destitution.
     Had he been interested only in the epic quality of the Oklahoma
to California exodus, Steinbeck would not have needed to go into the

field, to hear anecdotes and "complaints and . . . triumphs." The reports and photographs already in his possession would have been sufficient background material. I suspect he found himself too much in the position of Mac, the labor organizer of *In Dubious Battle,* someone who was schooled in theory but unable to penetrate "the mob." Throughout his life, Steinbeck was fascinated by what he called "phalanx man," the character of the group when it acted collectively, the analysis of which he thought was unequal to psychology of the kind used to predict individual behavior. But collective action is amenable to description from the outside. In being inside, in sitting in ditches and eating sow belly, he was in search of singularity; he was looking for the animating detail, the quality of felt experience, the press of history on private sensibility.

If he discovered the effects of history on the lives of itinerant farm workers, he also discovered the workings of power in another sense. Tractors, banking practices, and desolate weather might have sent Oklahomans out on the road, but they were not met in California by "the forces of history." Once there, their lives turned into a series of dodges from forces of another kind, from authority taken into the hands of a belligerent farming establishment. It is this concern of the novel that I would like to isolate, for it is here that Steinbeck continued and amplified the preoccupation with surveillance that had characterized *In Dubious Battle.* The sporadic roadside confrontations of the earlier book became, in *The Grapes of Wrath,* the signs of a condition that is general, the mark of a world that is convoluted in the way that Shakespearean tragedy is askew, with values obscured and violated until the legitimate heirs of power can reclaim their place.

This usurpation of legitimate authority is discussed most directly through Tom Joad, the loved and respected middle son of the Joad family. Steinbeck opens his novel with Joad's return from MacAlister Prison; this beginning, like Henry James's use of Milbank Prison in the opening of *The Princess Casamassima,* is emblematic of an entire world that is guarded and poised for disciplinary action. Pa Joad's first anxious question, "Tommy . . . you ain't busted out? You ain't got to hide?" (77), answered negatively, locates the central irony of the text. For both in Oklahoma and in California, it is not convicts who hide, but common people. *The Grapes of Wrath* is about a world that turns the ordinary into the fugitive; it is about an historical period that, in Steinbeck's eyes, inverted the very idea of imprisonment, so that Tom Joad's freedom is no release at all. His first experience on his father's farm is the need to hide from the representatives of the new landowners. The experience of the family as it travels in search of renewed life is a ritual reiteration of this episode: they become migrants because they have chosen to look for good work; once they are migrants, they are kept on the march, forced as much as any chain gang to keep moving. "I done it at Mac for four years," Tom says, "jus' marchin' in cell an' out cell an' in mess an' out mess. Jesus Christ, I thought it'd be somepin different when I come out!" (190).

This recreation of the American landscape as prison is Steinbeck's darkest musing; it is what makes the book, as the Nobel Prize committee noted, "serious and denunciatory." As counterpoint to Steinbeck's mythic sense of the power of common people to survive, a sense akin to Faulkner's claim for the endurance of blacks in *The Sound and the Fury,* is a very specific reconstruction of an historical situation where citizens live in constant fear of surveillance and where flight is the only mode of resisting "the eye of power." The very tentativeness of the migrants' housing, their patched canvas and cardboard dwellings, is a sign of an unseen power play, since movement can express both desire and evasion; it can be a response to elusive hopes or to experienced threats. This is what Dorothea Lange understood when she fixed her camera not on people working in wretched conditions but on people moving and living in the makeshift manner of the road. It is this perception that led her and Taylor to posit a solution to the migrant problems in terms of housing and government camps: in a world where the state has become a private police state, the only freedom is to be found in enclosure, in space that protects people from the vigilance of those who want to frighten them into quietness and submission.

In fact, these are exactly the terms in which the battles over the F.S.A. camps were fought. Charles Benson, in his superb investigation into the personal circumstances of Tom Collins, to whom *The Grapes of Wrath* is dedicated, quotes from the Yuba City *Herald* to show the extent of the anger directed toward these havens by the farming establishment:

> MIGRANT CAMP IS RED HOTBED
> The Federal migrant camp in the city of Maryville is becoming a Red hotbed, a breeding place for the fomenting of strikes to destroy the peach crops of Sutter, Yuba and Butte counties. The U.S. Government is sheltering these Reds at night, providing them with roofs, beds, and living accommodations, together with a willing audience.
>
> The Maryville City Government can either keep that migrant camp cleared of Reds or the ranchers will level it to the ground (9 July 1936).

Taylor had anticipated this kind of resistance to the camps in his S.E.R.A. reports, since he knew fences would permit the exercise of First Amendment rights. Steinbeck addressed the tension of the whole situation when he wrote for the *San Francisco News* in 1936, "A man herded about, surrounded by armed guards, starved and forced to live in filth loses his dignity; that is, he loses his valid position in regard to society, and consequently his whole ethics toward society. Nothing is a better example of this than the prison."

The Grapes of Wrath dramatizes both the anxiety of life on the road for the Joads and the relief of temporary stasis: the family moves to the Weedpatch camp through a nightmare of blockaded highways and invaded tent cities—an existence so haunted by armed guards that even communication among peers is distorted, talk suppressed, eye contact avoided. Steinbeck describes the release experienced by the family in terms of cleanliness and decency, in terms of sanitation and friendliness, but it is clear that the temporary health and expansiveness of the Joads arise from the walls tht shield them and that their chatter expresses their knowing escape from the "prison" outside. In a world where "spies [are] sent to catch the murmurings of revolt" (262), they have jumped bail.

The tentativeness of this resolution of anxiety in the novel reflects the tentativeness of the camps in actuality: there were over 200,000 homeless people in California, but only two housing projects were built, as experimental responses to Taylor's recommendations. These circumstances suited Steinbeck's purpose in writing: shelter of the kind offered by walls, the shelter of a fortress, could provide only temporary respite to a larger situation of siege. He understood the conflict in his book to be of a more fundamental order, however; he was interested in the endurance of values in the face of overwhelming disaster, in the resilience of spirit, in whatever caused people to define the limits of their own degradation. In the end, he-slid into a facile romanticism. The flood that ends *The Grapes of Wrath* leaves the Joad family at a moment of absolute poise between life and death: breast milk—which is all that is left of the material world for them to give—is offered by Rosasharn to an old man who is starving. It was a stark and extraordinary move, but in concluding the text in this way, Steinbeck violated the logic of his own political vision; he allowed a mythic folk worship to obscure the actual predicament of the folk to whom he had turned his attention. He careened off into a vision of generosity extended at the edge of darkness, of humanity asserted against the threat of extinction.

But to this point, *The Grapes of Wrath* had been a text created out of tensions that were human rather than natural in origin. He had pitted the will of the common worker against the more damaging will of the landowners and asked who would finally prevail. He had pointed out the futility of spies who caught "the murmuring of revolt" because "only the means to destroy revolt were considered while the causes for revolt went on" (262). The book was at base a dark vision of a society in radical disequilibrium; it created a world order whose legitimacy he continually questioned and whose values were carried quietly and covertly by ordinary people. Within the novel, their revolt represented the possibility of deposing an usurping power, of ending repression. While the flood drew the Joads into another dimension, it obscured the social nature of the conflict that had hitherto engulfed them.

It was finally the social implications of his own position in the world which evaded Steinbeck as well. Tom Collins, who was the executive of Taylor's and Lange's project, Weedpatch's camp director, and Steinbeck's collaborator, unexpectedly left a record of the fruit of Steinbeck's work. Spurred initially by Steinbeck's example and motivated finally by his own distress, he wrote his own version of the Oklahoma migration. It was a work intended to be fiction, but, curiously, Collins shifted from thinly veiled autobiography to actual reportage when he wrote the section that described his experiences with Steinbeck. Charles Benson speculates that Collins may have done this to enhance the market value of the manuscript, that he may have wanted to display his association with a famous author, but it is equally possible to see the press of truth behind the move, the need to drop artifice in order to leave evidence of important events.

Bringing in the Sheaves initially confirms what is known of Steinbeck's movements from other sources. It does not give dates, but it charts the writer's visit to Weedpatch and his later travel to Visalia, where the two men exhausted themselves in helping flood victims. Collins was unstintingly generous: he lived in a shed on the outskirts of the camp with Steinbeck, shared his food, his time, his speculations about "the slum." But his primary function was a passive one: to provide a cover so that the writer could wander, talk, and participate in camp committees and routines. When Steinbeck left him to finish writing *The Grapes of Wrath,* Collins was noticeably shaken; the moment showed unmistakably the disparity between their positions: what was a research trip for one was the substance of the other's life. In the wooden prose that characterized his narrative, Collins tried to rationalize their separation as a simple branching of a united effort:

> I pulled myself together. "Okay, John. If you think that is the right move, you just go ahead home. I believe what you have told me. After all, there was a time when you, yourself, were in need of the bare essentials of life. . . . Now that you are rich and your name is famous, I think I know what you mean when you say you have a job to do.'"[41]

But the break was final in many senses. Collins and Steinbeck never met again, except in the strange, elliptical circumstance that explains both the reception of *The Grapes of Wrath* among the migrants[42] and Collins's disappearance from the Farm Security Administration.

When it was finished, Steinbeck sent Collins several copies of his novel along with a letter that said, "I have spared nothing." In retrospect, Collins remembered being frightened by receiving the book, and he identified his fear as fundamentally different from the fears he was already accustomed to mastering. That the various farm organizations would denounce the novel was both predictable and manageable. But he was devastated in another unexpected way:

> When the very people among whom I had been living and working became infuriated over the book, that made things quite difficult. The cooperation I had so long enjoyed from my people became opposition, and suspicion was rampant among them. . . . I had the unusual experience of sitting for an hour in one of the assemblies while the chairman of the slum council berated me as a Judas. . . . "Ain't it so, Mr. Drake, that you done had a guy here with you for nigh two full weeks? And aint it so you done let him know all 'bout us people? And aint it so you done give him all of that rotten stuff he done writes about us?"[43]

Collins recognized immediately what was at stake in this confrontation, for it was not the nature of the portrayal that upset the workers, but the very fact of Steinbeck's disguised presence among them. Collins's collusion in that ruse had violated the very principles the camps had been established to uphold. The assembly asked "Drake" to leave the camp, comparing him explicitly to "labor spies" whom they had previously uncovered. Though the vote did not carry, Collins nonetheless understood his career there to be at an end.

> There was no sleep for me that night so I passed the long hours by finishing the reading of John's book—my death sentence—the termination of my work and my efforts with the rural poor—for their confidence in me had been LOST.
> And that death sentence had been written by John Steinbeck.[44]

The melodrama of Collins's recreation of this moment should not obscure the importance of his analysis. Dorothea Lange and Paul Taylor, who had worked to establish the government camps he was running, had worked in the interests of an open society, where the opinion of all ruled the conduct of all, where transparency of motive and action was a sufficient spur to self-regulation. In the rural California of their time, it was precisely this transparency that was threatened; it was the secretiveness of a power structure taken out of the hands of public accountability that Weedpatch stood feebly against. Its walls defined a space where the rural poor had privacy, freedom from surveillance, and freedom to define themselves.

Lange's camera can be seen as a metaphoric Weedpatch: within the field established between herself and her subjects lay all the principles of the migrant camp: if it interrupted privacy, it was invited to do so; looking was mutual; and images were gifts bestowed by knowing subjects. Her photography was both the result and the sponsor of social cohesion, the binder of wills. Her memories of this extraordinary period of her life were memories of connection and trust. Collins's memories were of hostility and suspicion. Steinbeck's memories are unrecorded, but as Collins came eventually to see, it was he who bore the brunt of Steinbeck's tactics, not Steinbeck himself.

Why a writer so obsessed with the damage of surveillance and spy tactics should have resorted to these strategies himself is not clear. But we know that Steinbeck throve on secrecy and that he understood his success as a writer to be predicated on anonymity. "Tell no one I am coming," he had written to Tom Collins. Ostensibly he was warning Collins not to let the Growers' Association know of his movements, but there was some covert desire expressed in this instruction as well. In a letter to Joseph Henry Jackson in 1935, in the midst of his involvement with the migrant labor scene, Steinbeck wrote, "In the last few books I have felt a curious richness as though my life had been multiplied through having become identified in a most real way with people who were not me."[45] He added that he feared fame might rob him of this ability. Later when offered the Commonwealth Club Award for *Tortilla Flat,* he reiterated the same anxiety about "cut[ting] [him]self off from their society forever."[46] For whatever personal reason, when writing Steinbeck needed to obliterate his ego, to become the "not me." He wanted to stand in the dark, behind the headlights, and look around. He considered this transaction a kind of richness and it is both appropriate and sad that he should use the language of gain to describe his feelings. For his search violated everything that his adversaries in the central California valleys had violated by their vigilante activities. All of them had secretly penetrated the walls that Weedpatch erected against surveillance and suppression. By entering and looking, he turned the experiences of the unwitting poor into private capital; by his stealth, he robbed them of the self-presentation that allows people to encounter each other as equals, and by taking, he destroyed their gifts.

Collins, like Lange, had known that generosity was also a mark of dignity, that coffee accepted or meals enjoyed created an enduring social fabric among people whose self respect was already battered. His usual way of dealing with the Oklahoma migrants was to accept food. By working with Steinbeck, he distorted these gifts and shifted his strategic position in a way that could not be redeemed. Where Lange and Taylor had built the camp he managed, he considered that Steinbeck tore it down. Acting as the writer's surrogate in punishment, he was cast out of Weedpatch, which had so hopefully and fragilely represented a new order of social organization.

Dorothea Lange. Camp manager, migratory worker, and visitor to the camp in discussion. Shafter (FSA) camp for migrants, California, 1938 *(Library of Congress)*

Dorothea Lange. Entering FSA camp for migratory laborers. Indio, Coachella Valley, California, 1939 *(Library of Congress)*

Dorothea Lange. The Arvin migratory farmworkers' camp of the FSA. Tom Collins, manager of the camp, talking with one of the members. Kern County, California, 1936 *(Library of Congress)*

Margaret Bourke-White. German Bunker outside Würzburg. 1945 *(Life magazine*
© *Time Inc.)*

7

Norman Mailer
and Combat Photography

Exposure under Fire

Or did they intuit, on a moment's reflection, that
war is the camera's ultimate challenge?
                                        —Wright Morris

I

YEARS after he had won acclaim as a young novelist of World
War II, Norman Mailer wrote a series of "advertisements" for
himself; ostensibly they were a publicity stunt, a contribution
to his life as a performance artist, another act of virtuosity in a career
already studded with "variations and postures." By 1959, he had
departed from the theme of war which had originally brought him
fame, written two subsequent novels about wildly different subjects, and
"sourly" declared he was destined to make a revolution in American
consciousness. Gaudy, provocative, taking pride in himself as a quick-
change artist, he chose nonetheless to describe his new posturing in the
world with the oldest and most profound images of his career:

> There is a time when an ambitious type should fight his way through
> the jungle and up the mountain—it is the time when experience is
> rich and you can learn more than you ever will again, but if it goes
> on too long, you wither from the high tension, you drop away drunk
> on a burned out brain . . . it is inevitable that a bad fall comes to the
> strong-willed man who is not strong enough to reach his own peak.[1]

233

Eleven years after the publication of *The Naked and the Dead,* the "jungle" and the "mountain" still expressed his sense of life's challenge and its threat. The mountain was the overview, the achieved mastery, the command position; the jungle was the dark zone of conflict in which sight was occluded, myopic, and solitary. Both had been scenes of actual battle in the South Pacific; both came to describe the parameters of his struggle as an artist. It was as if an originating, literal experience of combat never really ceased for Mailer but transformed itself into a recurring figurative language for identifying the wounding, aggression, and risk that continued to characterize his life and his imagination.

He came to these distinctions through experience. In one of the first interviews he granted after the success of *The Naked and the Dead,* Mailer spoke of his service in the South Pacific on Leyte and Luzon and later with the occupation forces in Japan. He was trained as a field artillery surveyor, assigned to an intelligence section of the 112th Cavalry, and later reassigned to interpret aerial photographs.[2] It was a pivotal move for him as a young novelist. Because of it, he came to see the "general's-eye view" of the war and to understand the fundamental discrepancy between war as a matter of strategy, tactics, and command, and war as experienced by the GI. The tensions between these two perspectives propelled the narrative he came eventually to write, and they also served as a way of speaking about more fundamental issues of American social organization that seemed, during the stress of combat, to be nakedly exposed to scrutiny. Nothing was isolated. As he put the matter within the text, "[T]he fact that you're holding the gun and the other man is not is no accident. It's a product of everything you've achieved."[3] It was also no accident, Mailer believed, that aerial surveillance photographs were the extended eye of power and that ground photography suffered such an abused fate during the war. These two uses of the camera represented two radically opposed ways of seeing and relating to the exercise of force; they were the vision of the mountain and the jungle made literal; and his task, in writing *The Naked and the Dead,* was not only to expose those structures and the values which empowered them, but also to place the writer—another kind of visionary—among them.

II

When Mailer arrived in the Philippines in late 1944, aerial surveillance had been in use for a little over three years. At the start of the war, the United States used no photographic intelligence at all. The many, small islands which became the arena of conflict in the Pacific were largely unknown even in their major contours. Before photography could serve more delicate strategic purposes, it was used simply to map the terrain, to disclose basic topographic features at the scenes of landing and of battle. Lieutenant Commander Robert S. Quakenbush had been the first officer to work out a program for Naval Photographic Intelligence, and his plans led to the establishment of a photographic group (commissioned on 1 October 1942) for collection and interpretation of material. These squadrons went to Espiritu Santo, to Guadalcanal, and then worked independently until 21 June 1943, when they were temporarily merged with army, navy, and marine groups known collectively as the Photo Wing South Pacific. This new composite group worked in tandem with the Navy's South Pacific Photographic Interpretation Unit and the 955th Engineer Topographic Unit, whose task was to exploit the imagery collected in the air. These cooperative missions continued in New Georgia (June 1943) and the Mariana Islands (April 1944) and were ended in June 1944, when it was judged that each branch of the service had enough resources to form its own, independent photoreconnaissance organization.[4] When Mailer's commanding officer in *The Naked and the Dead* looks over the surveillance photos of Anapopei, he is presumably receiving information from a squadron within the army's own command. When Mailer speaks of the general's "mental image of his battle map" (112), he is referring to a plan of action for ground troop movement that originated in the air, with aerial cameras.

George Goddard, who developed many of the lenses and techniques used in this kind of reconnaissance, remembered the armed services' initial resistance to the airborne camera. He considered that the pre-World War II armed forces had been primarily interested in force, not in surveillance, and that they needed to be shown the ways that the correlation of vision and power could enhance power. According to his autobiography, Goddard's life was fulfilled at the point when he was able to effect that full collaboration,[5] and his success, his effort on behalf of aerial surveillance, was eventually encoded in the army training manuals which explained to prospective photographers the importance of their work. Aerial photography had three primary functions: when photographers mapped local areas, they were contributing to decisions about future attack or patrol routes, bivouac sites, command posts, "radio relay sites, lay out wire or cable communication lines," and to an overall understanding of an area's topography. When they were sent over enemy territory, they were making images of its defenses. (What "troop concentration and movement, attack routes, ford and bridge sites, supply routes, artillery and other targets, and troop strength and

weapons" were visible?) Finally, they could be sent to photograph
friendly territory.[6]

Once these pictures were made, they needed to be decoded in a
series of increasingly complex ways. As one of the photo intelligence
instructors at Medmenham, England said, most of the photo interpreters of World War II were "bright young people who never used an
aerial photograph before 1940." This was certainly Mailer's position in
the Philippines, so he also faced what this same instructor identified as
the primary difficulty in learning to interpret aerial photography: "[I]t
entailed another view of the world—mainly looking down on it
vertically."[7]

But if this was a difficulty for Mailer, it was also an opportunity.
Though his work—and the work of all photo interpreters—entailed
meticulous decoding of scale, shapes, patterns, textures, and shadows,
these fragmentary, tedious calculations were the single most accepted
source of information among military strategists. To have a decision to
move troops or to bomb an enemy target "photo confirmed" was to
proceed with authenticity, objectivity, and reliability. The camera, as
used by the armed services in this way, was once again aligned with
positivistic science, with the rational view, and as Margaret Bourke-
White saw when she herself went up on an aerial mission, with an
aesthetic appreciation of the logic of strategy:

> As we headed toward the front I was impressed with how regular the
> pattern of war, seemingly so chaotic from the ground, appears from
> the air. The tracks of pattern bombing on an airfield were as regular
> as though drawn with ruler and compass. In some dire groves the
> traffic patterns made by trucks and jeeps which had parked there
> looked as if a school child had drawn circles in a penmanship exer-
> cise. . . . It was cruelly contradictory that with all this evidence of
> bloodshed and destruction, the valley seemed to clothe itself in a
> sequin-dotted dress.[8]

Mailer's General Cummings in *The Naked and the Dead* is this aes-
thetician of war, a commander of the heights, an amateur philosopher
of power. The categories of his thought are precisely those engendered
by his sources of information. Like the photographs that prompt his
decisions, the general reasons in terms of objectivity, pattern, and gen-
erality. "The idea of individual personality is just a hindrance," he tells
us. "I work with grosser techniques, common denominator techniques"
(181).

Hearn, the lieutenant to whom Cummings speaks, poses the central
challenge to this perspective: it is a challenge not only to a single gen-

eral, but to the underlying assumptions of his command, to his assertions of rationality and to the adequacy of the overview which separates
the universal from the particular and disregards the importance of
detail. Hearn finds the conversation unclean because of what it ignores:
"Somewhere out in the front a man might be rigid with terror, in his
foxhole." Speaking in behalf of the particular, he tells the general,
"You're so damn high you don't see anything at all" (181).

This dialogue both explains and embodies the narrative's central
dynamic. In its light, we can see the emplotment of the text as a serial
testing of the overview: at each strategic moment, the general knows
what aerial reconnaissance has given him to know:

> As if in assurance, the scattered air reconnaissance that was granted
> Cummings from army headquarters, brought back photographs
> which showed a powerful defense line set up by Toyaku on a front
> which ran from the main mountain range of Anapopei to the sea (46).

> Air reconnaissance had shown that fifty or perhaps even a hundred
> Japanese troops were entrenched in bunkers and pill boxes on that
> stretch of beach (381).

> He got out of bed, and trod over to the duckboards in his bare feet to
> examine some aerial photographs in his desk. Could a company do
> it? It was quite possible. . . . He studied the aerial maps (399).

The function of each episode that illustrates the general's response to
photographic intelligence is to discredit his assessment of what is the
case. He is shown to be wrong on the level of strategy and wrong in
the knowledge of its costs: aerial photography has located a line of
defense, but it has failed, entirely, to establish the strength of that
defense. Under actual attack, the Japanese prove to be so worn down
that "anyone could have won this campaign." The reconnaissance mission that Cummings sends through the gap in the Watamai mountains
near Mount Anaka to descend into the Japanese rear is demolished by
a totally inadequate appraisal of the terrain. Indeed, after the war, when
the U.S. Army evaluated the problems particular to air reconnaissance
in the South Pacific, it spoke specifically of the opacity of the jungle,
its resistance to evaluation in terms of the human strength required to
master it: "With the exception of a few operations in New Guinea and
the Philippines, ground engagements often tended to be close-gripped
struggles between small elements under a thick canopy of tropical
growth that hid the ground from the glass eye overhead."[9]

If Cummings's perspective is shown to be wrong, he is nonetheless
imbued with the power to be wrong. His vision prevails because he has
the authority to insist on it in the face of all resistance to it and even

in the face of his own self-doubt. Mailer shows cleanly and beautifully the fallacy of his objectivity; he insists that Cummings's facticity is used in the service of something far more sinister and dark and private. The general adheres to the aerial view not because it is without bias but because it is distant, solitary, and the means of control, even if control is of the internal structure of the army and only secondarily control or conquest of an enemy.

Mailer's general is, in this respect, the embodiment of a way of thinking, the personification of a naïve positivism, and an illustration of its status within a structure whose goals are domination rather than truth. Cummings's objectivity has no independent status; it is very simply a tool of policy, although policy is said to be made in deference to it. Put into the terms of the narrative itself, this means that Cummings's interest in the war is self-interest: he makes strategical decisions and evaluates combat only insofar as successful combat will promote his own career. His men are instrumental to this purpose, and this is what Hearn comes to see firsthand and what all of his troops sense intuitively. In fact, their response to the general is an extended instance of Foucault's observation that "there are no relations of power without resistances; the latter are all the more real and effective because they are formed right at the point where relations of power are exercised."[10]

If Mailer's narrative tests and rejects the efficacy of aerial surveillance, it also raises questions about alternative ways of understanding combat and the social structure exposed by it. The men's resistance to Cummings served as an index of domination and resentment, but Mailer was concerned to know what perspective could adequately survey the entire scene of conflict. The obvious alternative, that of the combatants themselves, he found inadequate, although it fascinated and lured him into self-created situations of risk in later life when the actual risks of war had abated. To be one of the men was to know war in a way that was denied by the loftiness of the command perspective; but to live and work among them was to be excluded from understanding the larger strategies which dictated the movements of individual lives. Mailer knew that the terrors of war came not only from the aggression of the enemy, but also from a sense of victimization within an inexorable machine, a sense of powerlessness within the iron purposes of one's own military organization. The men in Mailer's exemplary platoon often do not even know where they are:

"Where are we now?" Ridges asked.

"Second battalion headquarters," Croft said. "What've you been working on the road all this time if you don't even know where you are?"

"Shoot, Ah just work, Ah don' spend mah time lookin' around,"
Ridges said (263).

To "look around" was the specific task of the Army Signal Corps photographic units and of accredited pictorial correspondents from newspapers, magazines, and newsreel services. The position of these men in
combat was unprecedented, anomalous, and often handled badly. As the
army itself admitted in its own official history of the Signal Corps,
"Some few individuals saw the entire problem . . . but by and large, too
often through all ranks and grades the photographic mission was an
irritating gadfly—sometimes to be slapped down, more frequently to
be brushed away, and often merely to be ignored."[11]

The beginning years of the war were years of struggle to define the
role of the combat photographer and to find practical ways to integrate
the war with the record of its own conduct. To the extent that officers
could use ground photographers for strategic planning and intelligence,
they tended to be accommodating: if a photograph could show the enemy's tactical methods, demonstrate field conditions that required engineering attention, or illustrate the consequences of inept fighting techniques, it would be welcome. The Signal Corps always considered its
first duty to be providing this kind of information about combat and
field operations. Its primary audience was the command staff, not the
public or the more amorphous eyes of history. Only secondarily was
its task to record the war for nonparticipants.[12]

To the extent that war photography adhered to these priorities, it
continued to be objective and instrumental, the tool, once more, of military authority. But Mailer's interest in the photographer-at-war was
not in his utilitarian purpose; in fact, it was the possibility of a collision
between the photographer and a command structure that most interested him. He was fascinated by the combat cameraman's potentially
subversive position in battle, for images made at the scene of conflict
could record precisely those things commanders could overlook, precisely the things Mailer's lieutenant in The Naked and the Dead had in
mind when he found his superior's reasoning to be "unclean" and when
he observed that height or distance could occlude as well as enhance
vision.

What did the collapse of distance and the status of combatant do to
perspective? Exposed, vulnerable to attack, the combat cameraman
could photograph danger only by experiencing it. For him, Cartier-
Bresson's "decisive moment," the moment of perfect emotional collusion with a subject, was also a moment of decisive action which
required him to risk himself as a possible enemy target. It was this
originating circumstance that caused Mailer to attribute to combat pho-

tography an authenticity of an order wholly at odds with the rationality seen by military authority. The danger, the fear, the chaos which surrounded the photographer and which were his subject unquestionably provided a different kind of truth. Carl Mydans, the *Life* photographer who worked on Luzon at the same time Mailer was there, addressed all these issues in his memoirs when he admitted that "clear patterns of economic and national pressures" had perhaps been visible to generals or historians, but that he had seen only "sharp images . . . of ordinary souls caught in the convulsions of war and war's aftermath." "No great combat picture," he had added, "was ever made in a headquarter briefing. And no camera that was ever late for an assault was ever 'filled in' later by comrades in journalism or survivors of the action. The camera must always be there. And behind it there must always be a man's eye, and a soul."[13]

Other accounts of combat photography, whether by professional newsmen like Mydans or Army Signal Corps trainees, all spoke of exposure to fire as the prerequisite of the image, and of the special vulnerability of holding a camera instead of a gun on the front lines. Although in the same structural situation as a regular combatant, a cameraman was without defense. These accounts also routinely identified the times when the activity of recording action had to be forsaken for action itself, when the needs of wounded men or a weakened defense turned them from recorders into medics or fighters. Colonel M. E. Gillette of the Fifth Army described the situation of his cameramen with a fine sense of their ordeal:

> The military action has consisted for the most part of artillery duels, artillery concentrations and attacks at night. . . . Mountainous terrain, mines, booby traps, uncannily accurate enemy artillery fire and rain, fog, or cloud filled valleys have handicapped the movements of photographers or limited the use of the camera. Most of the mountain heights have been taken by the formula of artillery and mortar concentrations in the early evening, infantry advances at night, and with everyone in the forward positions carefully dug in or concealed during the daylight hours. . . . Any daytime movement, even on or near slopes in many cases, usually brought on artillery fire. . . . Pictorially, the front line areas are extremely quiet since there is no movement of importance to photograph and any movement by cameramen brings on enemy fire. . . . The photographers and cameramen on front line assignments have conscientiously attempted to obtain spectacular or exciting material, exposing themselves repeatedly, in unsuccessful efforts for the most part, to obtain such pictures mainly because such scenes did not exist.[14]

Gillette was an army regular, and his report was presumably made within a military structure of accountability which demanded to know why so few invigorating or picturesque images were produced at the front. Robert Capa, working for *Life* magazine in North Africa and Europe, spoke repeatedly of the same chances taken, the same dreariness, the same visual poverty: "This sort of photography was only for undertakers and I didn't like being one. If I was to share the funeral, I swore, I would have to share the procession.... [E]very day I took the same pictures of dust, smoke and death." "I dragged myself from mountain to mountain, taking pictures of mud, misery and death."[15]

For Capa, this dreariness and horror became the point of his photography; his images were committed to the common soldier who saw nothing picturesque about war. The lack of drama was itself a kind of eloquence that encoded typical but undistinguished and numbing obligations. They were addressed to an audience of similarly undistinguished people who did not make strategic decisions, but to whom war was also "done." In this respect, Capa was a photographer of the kind for which John Berger longed when he said that photographers must learn to think of themselves as serving those whose lives are involved in the events photographed.[16] Capa's allegiances grew from experiences shared with other soldiers. In a sense, the confraternity of exposure and danger promoted the bonding with the subject that Cartier-Bresson found essential to superb photography. As Capa himself said, "I managed to get some good pictures and showed how dreary and unspectacular fighting actually is. Scoops depend on luck and quick transmission.... But the soldier who looks at the shots of Troina, ten years from now in his home in Ohio, will be able to say, 'That's how it was.' "[17]

Capa's memoirs allow us to place him within the entire war effort. In the midst of a general logic of combat which was buttressed by aerial surveillance, he, as another kind of photographer, worked with different motives, different allegiances, and different material. His self-reflection, sparse as it is, adds to our understanding of the multiplicity of photographic activities during the war and also reveals the dialogue that photography conducted with itself as it participated in combat: it could provide the aesthetic view or the dreary and horrible one; it could be objective or personal; it could be logical or chaotic and sickening. George Goddard probably had dedicated his life to the first goal, Capa had given himself to the other; one activity made destruction more precise and effective, the other, in a sense, demonstrated the effectiveness of destruction on everybody, regardless of status as observer, soldier, or enemy. Capa knew he photographed in order to challenge the

very circumstances that brought his camera into use. Like the protagonist of Philip Caputo's *DelCorso's Gallery,* he photographed war to end
war, a circumstance that makes his own death in Vietnam in 1964 all
the more bitter.

These approaches to the problem of making images of battle provided Mailer with an aesthetic context in the Philippines. Though he
did not write essays about photographs, his work as a photo interpreter
placed him constantly in the midst of them, where he could see and
evaluate their various strategies for representing the jungles and mountains he faced in actual experience, with all their dense, attendant confusions and moments of clarity.

III

Mailer always insisted on the symbolic nature of *The Naked and the
Dead:* Anapopei island had never existed, the reconnaissance platoon had
been his own creation, the battle strategy also of his own devising.
There is no reason to challenge his disavowal of the "realistic documentary" tradition,[18] for, in whatever way we read the text, it begins
by offering a composite image of the campaign in the Southwest Pacific
as well as an embellishment of the conditions that met the 112th Cavalry on Leyte and Luzon. As Jan Valtin wrote in the history of the 24th
Division, "[T]here are more than seven thousand islands in the Philippines, and nearly three thousand of them do not even have a name."[19]

MacArthur's decision to invade Leyte in October 1944—two
months before the originally planned date—changed immediate combat conditions in ways that no one present seems ever to have forgotten:
it meant that men had to fight their way across the island in the
typhoon season, it meant that supporting engineers had inadequately
developed strategies for laying roads and airfields in the general ruin
and muck of the rains, and that units generally operated considerably
under full strength because of insufficient air support.[20] If Mailer, whose
division arrived on Leyte on 14 November, had no desire to replicate
the experience of any single platoon in his novel, it is nonetheless true
that his emblematic troops marched through weather, terrain, and strategic maneuvers that resembled the situation MacArthur had engi-
neered: the misery of Mailer's troops is analogous to the misery
recorded in *Children of Yesterday, The Avengers of Bataan,*[21] and any number of the privately written division histories about these two invasions.
Heat, rain, dysentery, jungle sores, slogging through mud, and exhaus-
tion are the backdrop of all these accounts, and their collective motive

was always to vivify the particular, to give resonance to the statistical perspective provided by official army histories and to substitute uncensored versions of experience for the more heroic combat vignettes that the army had endorsed during wartime. Both Mailer and his fellow war reporters could and did season their retrospective accounts of battle with stories of Americans who tortured prisoners, robbed Japanese corpses, and killed the Japanese wounded instead of nursing them.

To the extent that *The Naked and the Dead* is committed to the emplotment of this kind of detail, it is a generic war narrative, but the book is distinguished from others by the function of these details. We should notice that Mailer was not interested in individual instances in themselves, but as evidence of a prior, authenticating presence. In his eyes, knowledge of the fear and disgust of front line conflict was the prerequisite of narrating it; the artist at risk, the artist as combatant gave him a status within experience, made him commensurate with his subject, the true speaker of it.

Mailer was able to write from a position unmediated by other cultural interpretations of his position, and it was this structural situation or empirical precondition that made his experience analogous to the combat photographer who could not register an image without being under fire himself.

The authenticity of the narrative, conceived in this way, was itself a major obstacle and the source of one of the novel's major obfuscations. For the reality of the front line position was, as Capa and his colleagues repeatedly asserted, its dreariness, its lack of visual richness, its absence of opportunity for drama except in moments when drama was unrecordable: "I got all kinds of pictures of dust, pictures of smoke and of generals," Capa remarked, "but none of the tension of battle which I could feel and follow with my naked eyes."[22] To adhere only to the perspective of combatant would be to surrender the opportunity to demonstrate the nature of the structure which immobilized and blinded him in the first place. It was here that the literal situation of fighting in the Philippine jungles and the metaphorical implications of that struggle were joined in Mailer's mind. The inability to see the whole picture during combat ("No soldier in combat is able to follow the course of battle beyond the ken of his own squad, section or platoon, particularly in tropical terrain where a man often can see no further than the front sight of his rifle"[23]) became the base of the figure; the jungle was external and internal, a dark place of combat and a place of limited insight, the locus of unalterable and agonizing duty and the emblem of the ignorance which perpetuated those duties.

To remain within the common perspective, however authentic, was to fail to see and explain the context which gave meaning to each man's situation within the whole. If Mailer had denigrated aerial surveillance because it could not reach the essential and vivifying details of the ground, he turned equally from the standpoint of the combat photographer. Neither, by itself, could render the truth of war as he understood it. In this respect, *The Naked and the Dead* is an extraordinary narrative about the difficulty of its own construction, about the impossibility of satisfactory perspective, about the mutual exclusions of authority in an organizational sense and the authority of immediate experience. The novel uses detail to authenticate itself, but the strategic plan of the whole is derived from knowledge available only to those with something like aerial reconnaissance capabilities. Mailer could not have written this without using a command position similar to Cummings's. The novel is about the exclusiveness, the danger, the limitations of each kind of perspective, but the novelist, having denigrated the power of each model, usurps both for himself.

Dropping to the level of textual explication for a moment, we can see Mailer reenacting his own testing of narrative paradigms through the experience of Cummings's *aide de camp*. The heir to several factories, he rejects his father's money and values without finding a code that can replace the business ethic satisfactorily. He drifts until he drifts into the army, where Cummings becomes a more articulate parent figure and the struggle continues. Hearn accepts nothing—neither the general's authority, his own potential authority, nor the resentment of the enlisted men. To the extent that the book has a privileged center, Hearn provides it, witnessing those things that never fully claim his allegiance. Since Mailer has indicted both the command perspective and the combat view, this irresolution is an oblique virtue; but Mailer awards it a Pyrrhic victory: the fate of Hearn's déclassement is his own destruction.

This narrative resolution assumes added importance when we recognize Hearn as Mailer's equivalent of a photographic image interpreter—an enlisted man who, like Mailer's younger self, was drawn out of the line of battle and placed at the side of authority:

> And as long as Hearn remained with him, he could see the whole process from the inception of the thought to the tangible and immediate results the next day, the next month. That kind of knowledge was the hardest to obtain, the most concealed in everything Hearn had done in the past, and it intrigued him, it fascinated him (85).

Hearn stands at a radical divide, seeing more than ordinary men do, but unwilling to face the authority his own position carries with it.

Cummings compares himself to Mount Anaka, "both of them, from necessity, were bleak and solitary and alone, commanding the heights" (563). Hearn is equally bleak, equally alone, but without a similar rationale for his solitude. Cummings's self-appointed task is to force Hearn to acknowledge the necessity of command, and the narrative takes the lieutenant closer and closer to recognizing his affinity for it: "With a shock, [Hearn] realized the trace of contempt he was beginning to feel for an enlisted man" (168).

In fact he comes not only to see the attraction of power but also the power he always had but refused to name. It is as if he has transmuted his father's authoritarian mannerisms into an antiauthoritarian stance, mistaking a transcoded posture for a substantive change: "Maybe I have my father's stubbornness. The closest things, the dominant patterns are usually unanswerable" (341).

If we continue to understand Mailer's treatment of Hearn as a displaced form of self-reflection, we can see that the turns of his fate—his rebellion, its rationale, and its consequences—reflect Mailer's own struggle to understand his position as an observer within a society tightly structured by levels of authority. With whom, on the symbolic level, does the artist have most affinity? Through Hearn, he brings together questions about "seeing" and militarism; and because Mailer conflates the American military with the society that sponsors it, Hearn's dilemma joins together the more general issues of observation, aggression, and power in American culture at large. In retrospect, we can see that the irresolution that plagues Hearn—his attraction to authority and his anguish at exercising it; his sense of the radical blindness of authority and the genuine quality of frontline experience; his obsession with risk as the precondition of authenticity—was to remain with Mailer, transforming itself into a preoccupation with the hipster ("The White Negro") and still later with the hippie (*Armies of the Night*), but always recurring in patterns the origins of which we can recognize in this first battle experience and in his struggle to find an observational paradigm adequate to its description.

IV

The creative impasse expressed by Hearn's death in *The Naked and the Dead* was not one that Mailer could resolve with ease. As much as he may have denigrated his lieutenant, making him inadequate either to action or command, he was nonetheless Mailer's best proxy. In fact, one could say that Hearn's dilemma was simply renamed at subsequent points in Mailer's career, although the categories of choice remained constant. The social order present to Mailer's imagination seems to have ossified as a chain of command with a radical disparity between the governed and the governors which left the problems of mutual visibility unresolved. Hearn had tried to remain in a neutral position, but Mailer came increasingly to align himself with the subterranean world, the world of those who, like his fictional soldiers, remained "controlled, denied and starved."[24]

The reasons for this attempted allegiance are not difficult to find, for Mailer continued to be preoccupied with the problem of authenticity in art, and, once back in the U.S., saw the values of New York and Hollywood as transposed versions of the authoritarianism and concomitant blindness of the army. Corporate America came to be seen in terms of the military brass whose superiority implied, by definition, a distance and lack of discernment that were anathema to art. Hearn had said to Cummings, "You take a squad of men—what the hell do you know about what goes on in their heads?" (180). Almost twenty years later, when Mailer turned to using the camera himself, he explained his movie, *Wild 90,* as a corrective to an "Establishment" which could see no further than Cummings could see into the subterranean world that it supposedly controlled: "The Establishment will not begin to come its own half of the distance through the national gap until its knowledge of the real social life of that other, isolated and—what Washington will insist on calling—deprived world is accurate, rather than liberal, condescending, and over-programmatic."[25]

His film—ninety minutes of embarrassed and obscene improvisation by Mailer and some of his friends—was to go "half the distance through the national gap"; it was to use language, attitudes, and postures that would presumably offend "white collar workers and intellectual technicians of the communications industries" by offering them glimpses into the profane effects of power. Profanity is, Mailer said, "the most ineradicable measure of the potential violence of social class upon class, for no one swears so much as the proletariat when alone."[26]

Aping the proletariat in his obscenity, standing around without a script, trying to think of something to say in front of a hired movie camera, Mailer identified his motives as communication with the heights: *Wild 90* was a message to the General Cummingses of the country whose aerial perspective obscured the real qualities of life in the cultural trenches of America. In this context, abuse was a form of dialogue; it was a strange, civilian equivalent of Robert Capa's war photographs of mutilated bodies and death: shock them into seeing what's here. Capa had had grave misgivings about taking photographs, but he had struggled through the issues of motivation and procedure only to emerge with an even stronger conviction of the need for giving such pictures to the world. He justified the blasphemy of the images with an absolute clarity of purpose: of World War II in England, he remembered one incident in particular:

> The last man to leave the plane was the pilot. He seemed to be all right except for a slight gash on his forehead. I moved to get a close-up. He stopped midway and cried, "Are these the pictures you were waiting for, photographer?" I shut my camera and left for London without saying goodbye.
>
> On the train to London, with those successfully exposed rolls in my bag, I hated myself and my profession. This sort of photography was only for undertakers and I didn't like being one. If I was to share the funeral, I swore, I would have to share the procession.
>
> Next morning, after sleeping it over, I felt better. While shaving I had a conversation with myself about the incompatibility of being a reporter and hanging on to a tender soul at the same time. The pictures of the guys sitting around the airfield without the pictures of their being hurt and killed would have given the wrong impression. The pictures of the dead and wounded were the ones that would show people the real aspect of war and I was glad I had taken that one roll before I turned soppy.[27]

Capa would photograph in the name of truth, he would use himself as a conduit of horrors because he believed that revelation might end them. But he handled the issue of intrusion and of taking intimate pictures of agony by a kind of self-imposed ritual of worthiness: he must join his subjects in danger; exposing himself would justify the exposure of others. He could not be a passive witness of pain; he could not exploit suffering for his career. One could say that Capa joined in a kind of ceremonial bargaining with death in order to record death with impunity.

Mailer, I believe, entered into a similar ritual of worthiness with the people of the "cultureless and alienated bottom" who fascinated him. If one were going to expose them, one would have to be like them. One would have to face the enemy together. Pauline Kael's caustic remark about Mailer's conduct in *Wild 90*—"It must have taken a Harvard man many years of practice to achieve that low-life effect; he didn't acquire it just for the movie"[28]—was not an incidental or offhand observation. It recognized the heart of Mailer's convoluted and rough-shod delicacy, his reckoning of the "status" that should be accorded to the "controlled, denied and starved" people, to those who formed the underside of a domineering but otherwise obtuse culture. Like Diane

Arbus, Mailer found vitality in the lives of outcast people: they jived; they were "elite."[29] And the reason for their vitality was also the reason he chose to make a non-acted or "existential" movie: both street life and unrehearsed filming exposed their participants to risk: it was the correlation of danger and creativity that made for the good hustle and the authentic film. Hazard gave them all a common psychic territory. As he put it in *The Deer Park:* "The essence of spirit, he thought to himself, was to choose the thing which did not better one's position, but made it more perilous."[30]

When Mailer wrote in *Esquire* about his intentions for the film, he contrasted his work to both the Hollywood feature film and the "traditional" form of the documentary: neither had validity because of the ruses, the anxiety of self-consciousness evoked by the production set-ups. To act in a customary Hollywood sense was to perpetuate the constipated habits of the culture-at-large; to use a script was to borrow the "words of men who had had too much money and controlled too many things" or to imprison oneself in the "air of other people's habits, other people's defeats, boredom, quiet desperation, and muted icy self-destroying rage." The alternative, as he described it in life, provided the basis for an alternative in art. One must "live with death as immediate danger . . . divorce oneself from society . . . exist without roots . . . set out on that uncharted journey into the rebellious imperatives of the self."[31]

But we should note that these hatreds and drives—the resentment of a culture that supervised and demanded a stultifying conformity, and the need to escape from it—were emotions that had been evoked by military service at the beginning of Mailer's career, and his answer to general surveillance was a permutation of combat experience: the front-line foxhole was replaced by the urban street and then, in one of the most oddly inspired moves of his career, by the camera trained on unrehearsed and "naked" actors. To be in the midst of artillery barrage and to be in front of an unprogrammed camera were analogous experiences. In the repetitive structure of his imagination, the camera simply replaced the enemy gun of warfare so that the "danger" of the exposure guaranteed an authenticity of emotion that Mailer otherwise found so desperately hard to achieve. The photographer engendered an existential situation of the kind Mailer first experienced at war: "Existentialism deals with situations like love, sex, disaster and death, all those accidents and ultimates whose ends are by their nature indeterminate: you are in an existential situation when something important and/or unfamiliar is taking place and you do not know how it is going to turn out."[32]

And so Mailer gaped and stuttered and shuffled around before Pennebaker's hired camera, having spent $1500 to film the kind of mugging and drinking he and his friends had done previously in a bar after the theater. It had begun as a game, as a chance for grown men to mimic the Mafia and the tough-guy style of the Italian underworld. It ended with Mailer's realization that he had "deliver[ed] some old and close-to-forgotten experience which had been perhaps more obsessive than he realized."[33]

Pauline Kael thought that the heart of Mailer's experiment was a solution to the problem of awareness of the camera in documentaries; but I am convinced that his deepest obsession was still with "exposure under fire." Defenselessness before the camera elicited, for Mailer, a real source of action. It is all the more ironic that the obsessive stalking of psychic danger, that is, the very nature of exposure itself, should reveal the entire conceit of Mailer's search. For it is possible to consider that in making *Wild 90* he had simply reemerged with the same problem that had hounded him twenty years before in *The Naked and the Dead,* where the alienation of the observer could be overcome only in the face of an external enemy. That is, Hearn's problem of déclassement, his failure to establish any satisfactory social allegiances, could only be handled by artificial, or rather, by asocial means. He could find a place for himself among the living only when guns were literally trained on his platoon and he stood in those terrible and barren circumstances which imminent death will provide for us all.

Understanding Mailer's need to be "shot at" tells us something of the depth and the sadness of the central creative dilemma of his life. For if the only way to feel connected to the lives of one's subjects is to join them at the moment when the threat of death obliterates all human distinctions, one must feel very disconnected indeed, very distant, very caught, I would hazard, by a breadth of vision which leaves one "bleak and alone, commanding the heights." Mailer was more at home with the aerial view than he could easily admit. Instead he aligned himself with the photographic image interpreter, preserving, like Hearn, a sad kind of honor through his own continued alienation. Using the combat photographer as a paradigm both in life and in the construction of art, hiding from the knowledge that social distinctions persevere outside the leveling extremity of war, he remained committed to art that was embedded in radical action. He refrained from the raw use of power in life and tried to mitigate the symbolic power of art, where, as Fredric Jameson has reminded us, "the will to dominate [can] persevere intact."[34]

W. Eugene Smith. "Saipan, June 1944" *(Published in Life, © The heirs of W. Eugene Smith)*

Photographer unknown. In enemy country—Germany—one soldier standing by while the other films action. Use of the tripod was encouraged at all times; stressed when long-focus lenses were employed as on this PH 330. From Signal Corp Historical Project F-2b: *Combat Photography* (Washington, D.C., 1945).

Robert Capa. "Every day I took the same pictures of dust, smoke and death."
Tunisia, World War II (Robert Capa, *Images of War*, copyright © 1964)

Lucas Samaras. From *Samaras Album* (New York: Whitney Museum, 1971).

Conclusion

Even though I was there I could go no further.
—Maurice Blanchot

IMPASSE

WHY has the use of the camera tended so frequently to stir what is dark and predatory in the literary imagination? Had James Joyce reflected on photography as a mode of access to the world he might have described it in the way Stephen Dedalus described Pyrrhus to himself in *Ulysses:* "Yes, a disappointed bridge." A connection not made, a hunger not met, a desire which remains unsurfeited. These dissatisfactions are not inevitable; they have been overcome or avoided by photographers like Cartier-Bresson, Walker Evans, or Dorothea Lange, all of whom in their individual manners, discovered ways to engender parity and a harmony of purpose between the self and the Other. But it is not hard to see that this book is, in the end, about loss and about the attempted compensations of conscience in the face of personal creative emergency. Finally, it is about the misrepresentations of the Other that can occur as a result of distortions originating in the world inhabited by the artist in search of raw materials.

One way to unravel the meaning of this constellation of values might be found by observing that technology has its own logic. It is not that logic cannot be violated or expropriated for other purposes, but it tends to push in a certain direction which must be recognized in order to be countermanded. Unless a photographer actively and self-consciously modulates his or her position in relation to others, the camera sets up the world as something to be looked-at rather than as something to be experienced from-the-point-of-view-of another or as a reciprocal looking back and forth. This may seem to be an obvious point, but it is important to recognize the discontinuities, stratifications, and crediting of a single perspective that such a configuration of visibility implies, and conversely, to recognize the interrelationships, struggles, and dialogue between differing values that are suppressed by the disposition of the community into those-who-see and those-who-are-seen.

This disparity and privileging might be considered to have its origins in the way that the camera initially set up a series of expectations about the quality and reliability of its own observations. From the autotelic language that was easily and eagerly employed to describe the camera when it was first introduced into American culture (recall Samuel F. Morse explaining that daguerreotypes were "painted by Nature's self"), we know that it once seemed a fortunate way to suppress subjectivity, to avoid the distortions of idiosyncratic judgment and to affirm intersubjectively valid conclusions. It seemed, in short, to constitute one of the languages of objective reference, to be a form of empiricism. These associations, although later replaced by other associations with artistry and subjective expression, served as a kind of endorsement or accreditation of the photographer's truth.

I think it is correct to say that this early alliance with positivism initially obscured the dynamic nature of the face-to-face encounters or "conversations" which the camera promoted between photographer and subject. If nature inscribed itself on the photographic plate, if the photographer were present only as an agent of the sun, then one need not bother to understand or describe the variations of the photographer's engagement with his or her subject. One need not see observation as an exercise of power nor consider how the camera can function as a tool of cultural definition for those who hold it or act as cameraman. And yet, the struggle of many of the artists considered in this book has been to discover the reasons for their own uneasiness and for their troubled sense that spectatorship is not necessarily a neutral, abiding presence but a being-there which, because it is composed of balances and tensions and a struggle among competing points of view in a pluralistic culture, has an innately political character.

If we consider political action to be something collective and public, then this book charts a movement from one artist's solitary despair about his own internal depletion to a greater clarity of motive achieved by those who were able to reflect on the experiences of their creative precursors. In the 1850s, Nathaniel Hawthorne suffered quietly in the presence of a domestic subject; working privately and writing only about private life, he nonetheless came to understand that his creativity involved issues of desire, trespass, and self-imposed restraint. He came to feel that observation could serve as a means of power and as a mark of poverty, and that, in his own case, the two were not unrelated. The authority implicit in the creation of art often cloaked a need for control which grew out of isolation and the need to abridge pain by establishing a link, however tenuous or displaced, between the artist and the experiences of others. By the 1930s, James Agee could see that his entrance into the private lives of poor tenant farmers had ramifications far beyond his own personal discomfort. He understood that his behavior represented the attitudes of the news media and that it could, by extension, influence or shape policies affecting the lives of his subjects. In World War II, Norman Mailer was able to understand how camera surveillance could become the agent of the most terrifying military destruction.

As the camera moved into an increasingly central position in American life, each writer who reflected on its use and on its emblematic relation to his or her own creative practice became more aware of the implications of using social observation as a precondition of art. Each was increasingly self-conscious about what was involved in attempting to see and in being perceived, and each came to understand more forcefully that the experience of being the "targets of surveillance," to use Michel Foucault's harsh language, is at the heart of the experience of human powerlessness. As Foucault also points out, "[It] does not matter what motive animates [the one who stands in the central tower, who sees without ever being seen]: the curiosity of the indiscreet, the malice of a child, the thirst for knowledge of a philosopher who wishes to visit the museum of human nature, or the perversity of those who take pleasure in spying and punishing."[1] By the very authority of one's talent with camera or pen, by one's strategic position, one is implicated in a transaction which serves to empower the self, to buttress an elite position even in the very effort which may seem to constitute an act of opposition to it. According to this developmental scheme, the seeds of consternation have been present from the moment the camera was united in imagination with the craft of fiction writing, but the full meaning of this anxiety has only become apparent with the passage of time.

But if we think of political action in a more basic way, as a mode of being present in the world, as a constant reenactment of values through interaction, or as a constant searching for the principles that should govern our participation in the lives of others, then this book tells the story of an impasse, a creative obsession which has been met by various artists, with variations offered by their particular historical moments, repetitively and without resolution. It is a story about a series of frames which seem to dissolve into gestures of futility, about approaches that violate, about desires not satiated. In *Deceit, Desire and the Novel,* René Girard claims, "Great novels always spring from an obsession that has been transcended." He speaks about the self and the Other becoming one "in the miracle of the novel," and of the novel's promise that one can attain a "supereminent position" which allows one to understand his or her part in a universal identity.[2] But in the novels represented in this book, such unity remains elusive. Affective poverty seems most often to be the underside of formal achievement; and the novels themselves, when seen as narrative solutions to the authors' efforts to come to know others,[3] point to the broken nature of the attempts to overcome isolation and to the difficulty of achieving solidarity under those conditions. The personal drive to expropriate the world through vision, to be drawn into what one sees, to unite vision with fulfillment, is rarely satisfied, and that longing is even more rarely accompanied by insight about the experience of being the Other, the recipient of such unusual but compelling scrutiny.

I choose to look at the results of this research in the second manner. The gap between who is framed and the one who frames—the place which Jay Cantor calls "the space between"—is for every artist in each generation a place of political action in private life. It is here that he or she can choose the type of approach to social interaction that will generate the text. When it is a solitary gesture, it appears to be apolitical; when it takes part in the great movements of historical struggle or political upheaval like the Depression of the 1930s, the Spanish Civil War, or World War II, it seems to be better described by what Marxists call déclassement—the situation in which a person may rise above or get outside of the perspective common to his or her own economic or cultural circumstances. But for me, something about these issues remains reiterative and static despite the changes which seem to characterize specific, historically-bound choices.

It cannot, for example, be an accident that we find both Paul Theroux and Philip Caputo, two contemporary novelists who have written about the photographer as protagonist, driven to reiterate and deepen the anxieties that Hawthorne expressed so many years earlier in *The House of the Seven Gables.* One writes about a photographer's illicit love affair on Cape Cod, the other about the horrors of photojournalism in Vietnam and Beirut; one chooses domestic life as a fictional arena, the other the life of action in the world; but neither escapes the anxiety produced by visual predation. Hawthorne's burden as a hidden manager of lights and shadows was not, as it turned out, a solitary or idiosyncratic preoccupation; he was simply the first writer in our literary history to try to name the camera's most negative mode of engagement with the world and to equate that with his own predicament as a text-maker. He was the first to hit against the bones of a dilemma whose shape has only become clear through time and thwarted effort.

In *DelCorso's Gallery,* Caputo writes about a guilt so deep that it can only be expiated with the photographer's death. No less than Holgrave, Caputo's protagonist pushes against the walls of the unknown, driven first by obsession, by anger, by a vaguely understood need to make the American public pay for its complacency through the assault of magazine images of death and destruction. Working in Rach Giang, Vietnam, DelCorso had "gone in, pressing his shutter button as unthinkingly as the soldiers pulled their triggers, aiming his lenses as indiscriminately as they did their rifles. He'd shot the butchered family moments after they'd been cut down. Almost literally shot them."[4] This massacre seems to be a "monster got loose," something external to the self, a beast which ranges a world in which photographers have a special immunity. But like Henry James's "beast in the jungle," this monster stalks the psyche as well; and, indeed, it serves as a description of the psyche, of human darkness and private hysteria: "He felt no remorse at that moment, only a cruel satisfaction in the bloodshed." "Some of them were still alive when he ran into the hut; they died in the time it took him to adjust his setting to the dim light and compose the picture, so that, when he exposed the frame, it was as though he had murdered them with his camera."[5]

Caputo's photographer comes to recognize his own negative purposes, and the novel takes shape around his attempted expiation, around a personal sorcery which pushes him to repeat the situation of carnage, to find the correlative of his Vietnamese travesty in Beirut, as if one could, by repetition with change, cancel a previous sin or tally a private moral scorecard. In photographing the disemboweled and castrated bodies of Christian victims of the Mourabitoun, he finds his opportunity. And in fleeing from the P.L.O., he encounters his death. The compensation for prurience or for fame sought in the face of mortal suffering, Caputo says, is not to make a better photograph or to have a more compassionate attitude, but to experience a fate which brings one into the situation of the subject, joined to the Other by an understanding which stands entirely outside of the possibility of representing the content of that understanding.

Knowledge, to Caputo, is a knowledge about frontiers and barriers, about the attitudes, motives, and behaviors which keep people apart. It is also about the ramifications of entering the undefended houses of others, of violating the walls which represent psychic closure as well as literal protection for their inhabitants. In an uncanny reiteration of Hawthorne's wish that "chimneys could speak and betray . . . the secrets of all who . . . have assembled within," DelCorso works his own variation of a Paul Pry theme, hovering around bombarded streets, as if artillery had done the work imagination desired, cutting away walls, smashing brick: "An Ottoman-era villa that had taken a direct hit looked like a cutaway dollhouse smashed by a destructive child. An entire wall had given way, exposing the wreckage of the private little world it had once concealed. . . . The exposed interior of the house brought . . . nameless horror."[6] Like Hawthorne's nameless protagonist, he trespasses in the name of truth; and like his precursor, he faces a crisis when he realizes that the deepest truth of his position is not an objective naming of events, a mimesis, but a breach in civility, a violation of the immediate human community, which grows out of his own hunger, his own desire, his own grief and hostility in the face of that which is already lost.

Similarly, Paul Theroux's *Picture Palace* is a prolonged meditation on the relation between life and work, a relation that initially seems to be one of contraries, of mutual exclusion. Theroux's protagonist, Maud Coffin Pratt, the supposed contemporary of Margaret Bourke-White, Imogen Cunningham, and Dorothea Lange, faces a retrospective of her life's work. The opportunistic young curator, who searches out and catalogues the prints, works from the assumption that Pratt's life is somehow "in" the work; for him to assemble a show is to assemble the signs of the developing emotional life of the artist. Pratt herself believes there to be no connection between memory and the photographic images which seem to record memory. For her, "There had been no link between what I was and what I saw."[7]

The novel is a prolonged exploration of the appropriateness of these views. What is a "picture palace"? Is it a house of art or a house of memory? If it is a monument of achievement, does it express or eclipse the experiences from which it has been drawn? The curator is proven wrong by Pratt, for as he pulls the prints out of their storage space in the windmill near the house, Pratt, in revery, provides the story of the person behind the lens: where Frank sees access to exotic experience, Pratt remembers emotional emptiness; where he sees growing maturity, she remembers the static quality of life in hotel rooms, as if "I didn't exist." But since Pratt's most profound search had been not for fame or for the vicarious experience of the lives recorded by the lens, the achievement counts for nothing in her own eyes. Instead she had hoped to use her talent to win the love of her too-well-loved brother. With a shock that sends her into years of psychosomatic blindness, she learns that Orlando is sleeping with her sister. This moment of wild recognition is a moment she unthinkingly records with the camera: hoping to make love to Orlando, she succeeds only in making images of his love for another. Though she aspires to the status of subject, she remains solitary, excluded, on the other side.

Her photographs seem, then, to draw a line of demarcation between themselves and what lies outside them, but Theroux's deepest concern is to unmask the hostility that is implicit in the gesture of photographing this illicit love. It is a compensation for envy which only masquerades as dispassion. And its existence, as evidence of what the lovers want most desperately to conceal, is a participation in life, an act, which has lethal consequences: "Now I knew I had driven them into the sea. I had killed them with a picture."[8]

The importance of these two novels lies in the reiterative structure of the experiences recorded in them. Their emplotment, one might say, reinscribes the power relationships of antecedent texts: the paradigm that underlies them, no less than those of Hawthorne's "Sights from a Steeple" and Sadakichi Hartmann's "The Broken Plates," is that of an observer or overseer who, in collecting raw material in the world, perpetuates a model of domination which he or she might, in other circumstances, see through and arraign. In both of the contemporary narratives, the case is taken to the limit: the photographer is identified as a killer, a stalker of forbidden and destitute lives. But though this formation might be extreme when taken literally, it serves beautifully as a metaphor for what is the case in another sense: the maker of art is, potentially, the eclipser of the identity of others, whom he or she "kills" into art, fixing them for all time in the postures granted by his or her own imagination. The answer to "Whose version of the truth will prevail?" is, under the dynamic established by surreptitious surveillance, undoubtedly the artist's version, for he or she makes the subject of his or her scrutiny into a perpetual Other whose absence from cultural discourse as a speaking, autonomous being may not even be noticed since it already is represented in some way, in someone's eyes, with some authority.

Through the narrative paradigms that organize the imaginations of these writers, as they are elucidated by their photographic analogs, we can begin to understand the central problems of "approach" in this research: these texts expose a dilemma which they are generally incapable of resolving; their authority subordinates, incorporates, and potentially violates the intentions of those who are represented; they remind us, in the words of Fredric Jameson, "that within the symbolic power of art and culture the will to dominate perseveres intact."[9]

TURNING AWAY

If we look ahead, the issues of disequilibrium and domination that repeatedly plague the writers and photographers in this book can be seen as one kind of explanation for the many self-referential strategies which modernism later adopted. If a satisfactory approach to the Other seems to be an insoluble problem, if social observation so frequently violates dialogue, participation, or the self-definition of the subject, then one can choose to dispense with the Other. One can carry the impasse to its logical conclusion and turn one's work into a demonstration of one's own inability to find identity through relationship. One of the clearest examples of this is found in the writing of Samuel Beckett, whose protagonists drone on and on about their own existence so that existence itself becomes reduced to a matter of utterance: were the voice to cease, the speaker would die, that is, he would fail of a social identity since his only contact with that which is beyond and outside of himself is in the reception (not the genesis) of the narration.

That solitary turning inward and away from intersubjectivity finds its photographic counterpart in *Samaras Album,* a book of autopolaroids, autointerviews, and autobiography. "What frightens you?" Samaras asks himself. "The separation between me and the things I see." "Why are you making art?" "So that I can forget my separateness from everything else," he answers.[10] Remembering the discomfort at posing before the movie camera of a friend, thinking of his own lost dignity, he decides to be his own "Peeping Tom." "I use myself and therefore I don't have to go through all the extraneous kinds of relationship, like finding models and pretending artistic distance or finding some symbol of geometry."[11]

With knowledge of the dissymmetry between artist and subject, but with a continued need to control and a sense that making art is inextricably bound to that control, Samaras uses the autopolaroid; he sets up a scene for himself and then enters it before the timed exposure takes place. In this way, he can control his subject with impunity and, indeed, that discipline might be called a kind of rebirth, a parenting: "And my real self was the product—the polaroids." Turning from the damaging control of others to the scrutiny of the self, thinking that "the ideal way of using people is using them like clay,"[12] but considering such behavior to be out of the question, he joins the observation of himself with the presentation of himself; he abolishes the barrier of the frame by entering it.

Cindy Sherman, another contemporary photographer, has done something similar in turning the camera on herself, but with a variation that has even more disturbing ramifications then does Samaras's closed circle of narcissism. For behind Samaras's work lies the assumption that there is a self to know in the most intimate of ways: posing always in his own apartment, among familiar objects, often concentrating on parts of his own anatomy, his body contains and expresses individuality. Sherman's images, however, challenge even the most basic assumptions about the nature of human identity. Her work careens out past problems of authority between an observing self and an Other to questions about the very attributes which seem to provide the boundaries of individual existence. When Sherman parades before the lens of her own camera, shutter bulb in hand, what difference is there between the observing self and the performing self? What act of self-consciousness unites a self so objectified with the self that wants to know through observation? What significance can we attribute to such acting and transacting?

One way to sort out these issues is to notice the control that seems to characterize this solipsistic project. As the editor of her book remarks, Sherman directs herself entirely; she is actress, scenarist, cinematographer, costume designer, and makeup artist; the artifice is completely her own.[13] She says, in effect, "I will choose how you will see me. I will solve the problem of the participation of the subject by becoming the subject myself." Were these photographs designed to reveal character or mood, this interpretation might seem satisfactory. But the contrived nature of Sherman's imagery casts doubt on such a reading, for each frame shows Sherman dressed as someone else. The costumes and scenery are all clichéd; the fantasies are stale, secondhand, as if they were distant echoes of someone else's idea of the roles she should play. Who taught Sherman to assemble a blond wig with a kerchief ("Untitled Film Still No. 17, 1978") or to wear a low-cut blouse with high heels and anklets ("Untitled Film Still, No. 15, 1977")? Why does a housedress with an apron seem appropriate clothing in the corner of a room with a worn linoleum floor and a scuffed door ("Untitled Film Still No. 35, 1979")? Sherman's photographs ask us to think about how we learn how to appear and they imply that the culture has constructed us. It is as if the self were an illusion or a contrivance of something other, as if we were being written by the world, authored by a diffuse circumference, as if the very idea of a creative and creating center were itself a fantasy. In this way, work that initially seems to be solipsistic or auto-deictic, turns from self-reference to emptiness, from active gesture to a form of receptivity. Issues of power and relationship recede into issues of usurpation because the artist, as subject, seems to be a dream that is dictated by a culture that cannot be represented in itself but only pointed to by its affects. In Sherman's world, a world which does not attempt to include others, even an approach to one's own being remains a problem of regression without closure.

In short, Sherman's retreat into selfhood ends by assuming a kind of negative political value. It serves as a demonstration of the fragility of identity even when gestures of self-definition seem to be most explicit. If this solipsism addresses the problem of art's symbolic domination of others, it does so at great expense, ending in an activity which calls attention to its own aridity.

CONFLUENCE

If this refusal of the world, if turning inward and away from dynamic conversation is one of art's responses to the difficulty of approaching others in a pluralistic culture, it is not the only way of coming to know and to represent the discourse of those who are unknown. Connections are sometimes made, multiplicity accommodated, others voices heard and acknowledged. But these bridges have rarely been built from the observations of a unitary and authoritative center; they have rarely proceeded from the exclusive position that the camera seems so superbly to encourage and to describe. In the two cases that I want to consider in closing—those of John Berger and Eudora Welty—photography, as an activity of looking-at and gathering-in, no longer serves as an adequate way to identify the writer's mode of entering into and embracing the world. In neither example is this an accidental renunciation, for each writer worked with or as a photographer; each knew from the deepest meditation that his or her craft proceeded from sources that originated in vision but left vision behind as sight was joined to insight, as observation gave way to identification.

Eudora Welty, who worked as a junior publicity agent for the Works Progress Administration during the Depression, knew especially what she had learned from her experiences behind the lens and what remained still to be learned when the camera was put down and the pen taken in hand. "The direction my mind took," she reflected, "was a writer's direction from the start, not a photographer's or a recorder's."[14]

When she recalled her early experiences with the W.P.A., she remembered the photographs themselves, for they provided a record both of her own past and of a time of desolation in the American South. But for her, the activity of photographing was the great teacher: "[M]ost of what I learned for myself came right at the time and directly out of the *taking* of the pictures. The camera was a hand-held auxiliary of wanting-to-know."[15]

It was a need, a desire almost as old as she was. As a child, she had squared her hands over her eyes, looking out of this box in the same way she had previously looked out of train windows. Framing, separating out, waiting for the moments when the world would reveal itself in gesture, yield up its secret, turn concealment to communication, she came to understand the way in which the outside world was, as she put it, the secret sharer of her own inner sensibility. Learning to see was one of her first lessons, but it was only a prelude of a more complex narrative mode. "This is not," she wrote in *One Writer's Beginnings,* "an observer's story. The tableau discovered through the young girl's framing hands is unwelcome realism."

> How can she accommodate the existence of this view to the dream of love, which she carried already inside her? Amorphous and tender, from now on it will have to remain hidden, her own secret imagining. The frame only raises the question of the vision.[16]

"How will I join vision with my own interior life?" is the question which will animate and propel her story-telling, and this joining, which is an interior action, will form, in a sense, both the substance and the subject of her narration. The framed world remains as a corrective to imagination, limiting the reaches of "vision, dream, illusion, hallucination, obsession and that most wonderful interior vision which is memory." But it is not, in itself, more than a reference or a reminder of the configuration of the spirit. The unseen world of human connection can be represented by means of "unwanted realism," but the visibility of the world is expropriated as one side of a metaphor which expresses the unity of perceiver and subject. The frame is not like a viewfinder or a small girl's lifted hands, but an unseen circle that enfolds everything seen and unseen.

This is, essentially, the vision which animates the conclusion of *The Optimist's Daughter* and which Welty uses again to conclude her memoirs. The woman in the novel dreams a dream that is also a memory: she is riding in a train from Chicago to Mount Salus with the man she will marry. At a bridge outside Cairo, she looks down at the juncture of the great Mississippi river and the Ohio and then looks up at a ragged line of migrating birds. At this moment, the visible world, glimpsed at a threshold, has the same contours as the lives of the two fervent lovers: the bridge, the converging rivers, the birds flying into the unknown are, at once, themselves, a vision, and an expression: "All they could see was sky, water, birds, light, and confluence. It was the whole morning world. And they themselves were a part of the confluence."[17]

And then, as if this lucid moment were not enough, Welty pulls her protagonist outside of time, and we learn that this is the memory of a widow, the thought of a woman whose previous certainty of immortality has been dashed by the death "in a year long gone" of her husband. But here, time is not the enemy; it is the great binder. The widow understands her own solitary existence to be a continuation of love, and Welty uses this moment of revery in the fiction to describe her own life of creativity outside of the fiction, to speak of her dearest resource, her means of connection to "the old and the young, the past and the present, the living and the dead." "Of course," she tells us, "the greatest confluence of all is that which makes up the human memory— the individual human memory."[18]

For John Berger, too, it is the photographer's ability to act like memory that forms the basis of his continuing interest in the camera. "If there is a narrative form intrinsic to still photography, it will search for what happened, as memories or reflections do."[19] To the extent that the camera remains a tool of positivism, he, like Welty, is disappointed in its use. In its traditional social observation mode, photography reduces truth to fact and implicitly suppresses "the social function of subjectivity." It can, Berger says, only "narrate descriptively from the outsider's point of view." "[T]he story told is finally about what the photographer saw at Y. It is not directly about the experience of those living the event in Y."[20]

The "experience of those living the event" is, by contrast, Berger's most central concern; and in his reflections about how to represent it, he comes upon the thought that it is no accident that photographers, as purveyors of fact, have rarely found the means of conveying the self-knowledge of their subjects. "Positivism and the camera and sociology grew up together," he observes. All of these modes of discourse have in common the belief that objective knowledge can order both the natural and social worlds, and in that desire to order and control is implicit the assumption that subjective experience has no place in the public sphere. It is devalued, or as Foucault would say, "discredited," cast low in a hierarchy of possible modes of knowing because it has no useful contribution to the tasks of planning and management.

But what if one were convinced, as Berger is, that the experiences of common, normally inarticulate people are valuable, and, indeed, that they constitute a treasure which is passing out of the world for want of a way to notice and preserve those experiences? One might begin to search out an alternative mode of representation which is not an eye-witness account but more akin to the raw material of memory which gathers within itself the events and experiences of different times and then works to bind them together. Memory itself would be the invisible protoganist; she would be "manifest in each connection made. One might say that she is defined by the way she wears the world."[21] To make the exercise possible, the photographer would have to stand in the place of the protagonist, looking out, seeing from-the-perspective-of; he or she would have to leave the position of the viewer-of-the-Other and move into the life experience of the Other. Inevitably, as in life, as in memory, the narrative would be discontinuous, a montage of images that would demand the activity of story-telling to provide coherence. The connection between the image of knitting hands and the image of a man in an undershirt holding a frisky colt might, for example, be the bonds of kinship, affection, need, and self-sufficiency in meeting needs: the winter is approaching; a farm woman is providing for her son or husband.

Berger's proposal does not forsake photography, but it does recast the nature of its habitual dynamic, where the cameraman's place of observation becomes the center of the world-as-it-is-experienced by the Other. What is unpresentable—the mind of the perceiving subject—nonetheless provides the fulcrum for a seemingly random array of images. In it is contained the memories which bind up disparities and allow a narrative to be generated.

This is a radical shift in the symbolic position of the artist vis-à-vis his or her subject. It reverses the logic of the panopticon where super-vision was implicitly allied with control and with its targets' loss of freedom. Berger undoubtedly understood and endorsed the subversive nature of this realignment of power, but it is important for us to note that these generative and regenerative actions occur in the mind, in the interior world. Like Welty, Berger locates the possibility of unity and coherence of perspective between people outside the world of time and effective political action. Memory, not history, is the binder of wills.

I have emphasized the role of memory in Berger's and Welty's reflections about creativity in order to call attention to the vulnerability of their solutions to the problem of approaching others. While they each seem to offer alternatives to the solipsism and refusals of interac-tion that characterize the work of Samaras and Sherman, they turn, in the end, to private and internal gestures for the coherence, equality, and reciprocity that they do not find in the historical world. In doing this, they reassert the inherent value of subjective appraisal, but they also leave us with some unsettling reflections about the public dimension of the practice of art.

If we think of the photographer as did Oliver Wendell Holmes during the first bright, optimistic years of photography's development in America, as a "mirror with a memory," then we are led to ask, "What was mirrored?" "What was recollected?" Holmes devised this figure of speech to describe the ability of emulsions to fix an image on a piece of metal or paper; he was thinking about the print's ability to hold and keep the appearances which in life itself are perishable and unstable. If we think of the history of photography in America as a history of prints, if we constitute history from the chemical processes and aesthetic principles which have governed the genesis of these prints, then we will find one sort of answer to these questions. Richard Rudisill, for example, in an excellent study called *Mirror Image: The Influence of the Daguerreotype on American Society,* summarizes "the medium's function in three primary ways—it served to cultivate nationalism, it helped Americans adjust themselves intuitively to the transition from an agrarian to a technological society, and it was ultimately a reflection of spiritual concerns motivating the nation."[22] He generalizes from the interaction of images (things) and people, and sees the democratizing influence of a form of representation that could, unlike painted portraits, represent all classes and types of people in affordable ways: one did not need to be wealthy to buy a daguerreotype, and one could, through the easy availability of images, find out what America looked like. Photography was both a form of mechanical ingenuity and a recorder of the other types of inventiveness that were changing the nation from a pastoral to an industrialized landscape.

But if, when we ask "What was mirrored?" we mean "What manner of commerce or human interaction was generated by taking photographs?", the answer will occur in quite different terms and in terms which challenge or show the fragility of the democratic principles which Rudisill—to use only one example—thought to be fostered by the camera.

This book cannot properly be called a history because it neither generates a continuous narrative nor claims to be inclusive. It proceeds by case studies. But to the extent that one can generalize from numer-examples, it leads us to see a disturbing pattern of American experience, a pattern in which Rousseau's ideal of the state as a transparent community of mutually visible citizens and motives is rarely achieved.

Alexander de Tocqueville's Rousseauesque characterization of America as a nation would rarely describe the "nation" of photographers, writers, and subjects whose confrontations have been discussed in these pages. Instead, these artists have tended, in their "decisive moments," to organize the visible world into relationships of domination and subjugation and to encode that inequity of generated meaning in their work. "It should be noted," M. M. Bakhtin tells us, "that the [work of art] always includes in itself the activity of coming to know another's word, a coming to knowledge whose process is represented in [the work of art]."[23] What is mirrored by photography and by the writers who have been influenced by it is, in this sense, American culture's tendency to organize itself hierarchically and to cloak interested motives in the language of equity while the true structure of privilege remains unnamed.

This is a generalization that does not describe every case, but it does locate a proclivity whose existence is further verified by the guilt that artists have repeatedly experienced as "seeking fabulists," as people whose activity of coming to know others is troublesome even to themselves. But their dis-ease has often been part of a struggle for greater self-reflection that, even when it is not entirely successful, allows us to see the damage a culture's unequal distribution of visibility can promote. Their example allows us to begin another struggle for the self-reflection that must buttress any truly reciprocal interaction in the world, a world whose multiplicity of perspectives is too compelling to be ignored and too important to remain unseen.

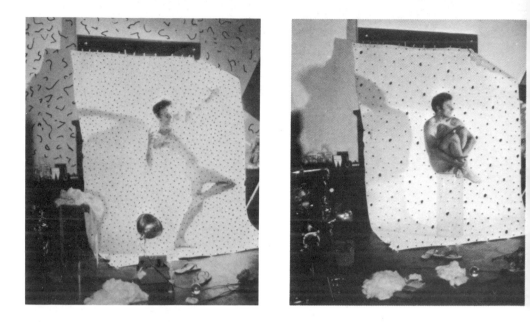

Lucas Samaras. From *Samaras Album* (New York: Whitney Museum, 1971).

Cindy Sherman,
Untitled Film Still No. 15.
1978. From *Cindy Sherman* (New York:
Pantheon, 1984).

Cindy Sherman. Untitled Film Still No. 35. 1979. From *Cindy Sherman* (New York: Pantheon, 1984).

Jean Mohr. Knitting Hands. From John Berger, *Another Way of Telling* (New York: Pantheon, 1982).

NOTES

Introduction

1. *Traité pratique de photographie* (Paris: J. J. Dubochet, 1844), quoted in Richard Rudisill, *Mirror Image: The Influence of the Daguerreotype on American Society* (Albuquerque: Univ. of New Mexico Press, 1971), 48.

2. Fredric Jameson, *The Political Unconscious* (Ithaca: Cornell Univ. Press, 1981), 182.

3. John Berger, *Pig Earth* (New York: Pantheon, 1979), 6.

4. *Ibid.,* 56, 55.

5. *Ibid.,* 9.

6. Theodor Adorno, *Negative Dialectics,* trans. E. B. Ashton (New York: Seabury, 1973), 17–18.

7. Henri Cartier-Bresson, "The Decisive Moment," in *Photographers on Photography,* ed. Nathan Lyons (Englewood Cliffs, N.J.: Prentice-Hall, 1966), 51.

8. James Agee, *Let Us Now Praise Famous Men* (Boston: Houghton Mifflin, 1960), 42.

9. Susan Sontag, *On Photography* (New York: Farrar, Straus and Giroux, 1977), 155.

10. Agee, *Let Us Now Praise Famous Men,* 232.

11. James Joyce, *Stephen Hero* (New York: New Directions, 1963), 78.

12. Edward Weston, *The Daybooks of Edward Weston,* ed. Nancy Newhall (Rochester, N.Y.: George Eastman House, 1961), I: 69.

13. Janet Malcolm, *Diana and Nikon* (Boston: David R. Godine, 1980), 133.

14. *Ibid.*

15. Robert Frank, "Statement, 1958," in *Photography in Print,* ed. Vicki Goldberg (New York: Simon and Schuster, 1981), 401.

16. John Le Carré, "Introduction," *Hearts of Darkness,* by Don McCullin (New York: Knopf, 1981), 17, 15.

17. Diane Arbus, *Diane Arbus* (Millerton, N.Y.: Aperture, 1972), 1.

18. *Ibid.,* 6.

19. Jean Jacques Rousseau, *The Social Contract* (London: Swan Sonnenschien, 1898), 126.

20. Brian Fay, *Social Theory and Political Practice* (London: Allen and Unwin, 1975), 54.

21. John Berger, *About Looking* (New York: Pantheon, 1980), 46.

22. See Jürgen Habermas, *Knowledge and Human Interest,* trans. Jeremy J. Shapiro (Boston: Beacon, 1971).

23. Walker Evans, *Walker Evans at Work* (New York: Harper and Row, 1982), 160.

24. See Michel Foucault, *Discipline and Punish,* trans. Alan Sheridan (New York: Vintage, 1979), 170ff.

25. Henry James, "The Aspern Papers," in *The Art of the Novel* (New York: Scribner's, 1934), 163.

26. James Agee, *Let Us Now Praise Famous Men,* 11.

27. Nathaniel Hawthorne, *The House of the Seven Gables* (New York: Dodd, Mead, 1950), 93.

28. Michel Foucault, "Two Lectures," in *Power/Knowledge,* trans. Colin Gordon, Leo Marshall, John Mepham, and Kate Soper (New York: Scribner's, 1980), 329.

29. Henry James, "The Golden Bowl," in *The Art of the Novel* (New York: Scribner's, 1934), 327.

30. William Carlos Williams, *Paterson* (New York: New Directions, 1963), 113.

31. Emile Zola, "The Experimental Novel," in *Documents of Modern Literary Realism,* ed. George J. Becker (Princeton: Princeton Univ. Press, 1963), 163.

32. Frank Norris quoted in Malcolm Cowley, "A Natural History of American Naturalism," in *Documents of Modern Literary Realism,* 442.

33. Quoted in Charles Child Walcutt, *American Naturalism: A Divided Stream* (Minneapolis: Univ. of Minnesota Press, 1956), 89.

34. James F. Ryder, *Voightlander and I in Pursuit of Shadow Catching* (Cleveland: Cleveland Printing and Publishing, 1902), 16.

35. Theodore Dreiser, *Newspaper Days* (New York: Horace Liveright, 1922), 454.

36. John Dos Passos to Dudley Poore, 15 Sept. 1916, in Townsend Ludington, *John Dos Passos: A Twentieth Century Odyssey* (New York: Dutton, 1980), 90.

37. John Dos Passos, "Vacancy," in *Century's Ebb* (Boston: Gambit, 1975), 348.

38. Margaret Bourke-White, *Purple Heart Valley: A Combat Chronicle of the War in Italy* (New York: Simon and Schuster, 1944), 5.

39. Robert Capa, *Images of War* (New York: Paragraphic Books, 1964), 78.

40. John Steinbeck, *A Russian Journal with Pictures by Robert Capa* (New York: Viking, 1948), 5.

41. Jay Cantor, *The Space Between: Literature and Politics* (Baltimore: Johns Hopkins Univ. Press, 1981), 12.

Chapter 1. Nathaniel Hawthorne and Daguerreotypy

1. Nathaniel Hawthorne to James T. Fields, 19 July 1860, MS, St. Lawrence University Library.

2. Nathaniel Hawthorne to William D. Ticknor, 23 May 1856, MS, Berg Collection, New York Public Library.

3. Nathaniel Hawthorne to William D. Ticknor, 5 June 1857, MS, Berg Collection, NYPL.

4. Nathaniel Hawthorne to William D. Ticknor, 23 May 1856, MS, Berg Collection, NYPL.

5. Nathaniel Hawthorne in *American Renaissance* by F. O. Matthiessen (New York: Oxford Univ. Press, 1941), 234.

6. For a list of existing photographs of Hawthorne see Rita K. Gollin, *Portraits of Nathaniel Hawthorne* (DeKalb, Ill.: Northern Illinois Univ. Press, 1983). I want to thank Rita for the clarification about the circumstances of composing Hawthorne's portraits that her book provided me.

7. Nathaniel Hawthorne, "Sights from a Steeple," in *Twice Told Tales* (New York: Dutton, 1961), 139.

8. *Ibid.*

9. *Ibid.,* 143.

10. François Gouraud to Henry W. Longfellow, March 1840, MS Am 1340.2, Houghton Library, Harvard University.

11. *Boston Evening Transcript,* 15 April 1840, p. 3.

12. Robert Sobieszek, "Introduction," *The Spirit of Fact: The Daguerreotypes of Southworth and Hawes 1843-1862* (Rochester: International Museum of Photography, 1976), v.

13. Beaumont Newhall, *The Daguerreotype in America* (New York: Duell, Sloan and Pearce, 1961).

14. Robert Taft, *Photography and the American Scene: A Social History 1839-1889* (New York: Dover, 1964). See also Pamela Hoyle, *The Development of Photography in Boston, 1840-1875* (Boston: Boston Athenaeum, 1978).

15. Edward Everett Hale, 15 April 1840, MS, Massachusetts Historical Society Archives.

16. Edward Everett Hale, *Journal II,* 31 March 1840 & passim, MS, Massachusetts Historical Society Archives.

17. François Gouraud, *Description of the Daguerreotype Process* (Boston: Dutton and Wentworth, 1840).

18. *Ibid.*

19. [Nathaniel P. Willis], "The Pencil of Nature," *The Corsair* I, No. 5 (April 1839): 71.

20. William Henry Fox Talbot, *The Pencil of Nature* (London: 1844; rpt. *Image,* Vol. II (June 1959).

21. Samuel Morse to the Editor of the *New York Observer,* 27 April 1839 in *The Life of Samuel Morse* by Samuel I. Prime (New York: D. Appleton, 1875), 400.

22. Jürgen Habermas, *Knowledge and Human Interest,* trans. Jeremy J. Shapiro (Boston: Beacon Press, 1971), discusses this issue further. His book provides much of the theory underlying this text.

23. Anson Clark, "Daguerreotype Portraits," in *Mirror Image* by Richard Rudisill (Albuquerque: Univ. of New Mexico Press, 1971), 126.

24. Marcus A. Root, *The Camera and the Pencil* (Philadelphia: Lippincott, 1864), 124-26.

25. Annie Fields, *Authors and Friends* (Boston: Houghton Mifflin, 1897), 105.

26. Richard Rudisill, *Mirror Image,* 126.

27. Ralph W. Emerson, "Nature," in *The Selected Writings of Ralph Waldo Emerson,* ed. Brooks Atkinson (New York: Modern Library, 1950), 6.

28. Edgar Allan Poe concurred: "In truth the daguerreotype plate is infinitely more accurate than any painting by human hands ... the closest scrutiny of the photographic drawing discloses only a more absolute truth, more perfect identity of aspect with the thing represented." In Rudisill, *Mirror Image,* 54.

29. James F. Ryder, *Voightlander and I in Pursuit of Shadow Catching* (Cleveland: Cleveland Printing and Publishing, 1902), 16.

30. *Ibid.,* xi.

31. John Fitzgibbon, "The 'Arkansaw Traveler' Daguerreotyped," *Photographic and Fine Art Journal* VII, No. 2 (Nov. 1854): 325.

32. Oliver Wendell Holmes, "The Stereoscope and the Stereograph," *Atlantic Monthly* III (June 1859): 162.

33. Henry David Thoreau, *Journal,* ed. Bradford Torrey (Boston: Houghton Mifflin, 1906), I:189.

34. Ralph Waldo Emerson, *Journal,* eds. Edward Waldo Emerson and Waldo Emerson Forbes (Boston: Houghton Mifflin, 1912), VI: 87.

35. *Ibid.,* 111.

36. See also Carol Shloss, "Oliver Wendell Holmes as an Amateur Photographer," *History of Photography* V, No. 2 (April 1981).

37. Oliver Wendell Holmes to Nathaniel Hawthorne, 9 April 1851, in George P. Lathrop, *A Study of Hawthorne* (Boston, 1876), 232.

38. T. W. Higginson, *Letters and Journals of T. W. Higginson 1846-1906,* ed. M. T. Higginson, 2nd ed. (Boston: Houghton Mifflin, 1921; rpt. New York: Negro Universities Press, 1969), 97.

39. T. S. Arthur, "The Daguerreotypist," *Godey's Lady's Book* (May 1849): 352.

40. Oliver Wendell Holmes, "The Stereoscope," 102.

41. Bernhard Freiherr von Tauchnitz to Nathaniel Hawthorne, 9 Jan. 1864, MS, Houghton Library, Harvard University.

42. Lucien Goldschmidt and Weston J. Naef, *The Truthful Lens: A Survey of the Photographically Illustrated Book 1844-1914* (New York: Grolier Club, 1980), 204. This information comes from C. E. Fraser Clark's unpublished research.

43. Rita Gollin's research into the controversies surrounding this portrait suggests that Hawthorne may even have denied sitting for this picture. She presents various confusing accounts from the memories of Elizabeth Peabody, Julian Hawthorne, and John Motley, but concludes (in agreement with Julian Hawthorne) that Hawthorne had this photo made in the company of Henry Bright on 19 May 1860. The original portrait was a large cabinet size of 6½″ × 9″ that was not intended for mass production. After Hawthorne's death in 1864, the Mayall studio issued copies of this pose in *carte de visite* form (2¼″ × 3½″).

44. Nathaniel Hawthorne, *The House of the Seven Gables* (Columbus: Ohio State Univ. Press, 1970), xiv. All subsequent references to this book noted by page number in the text.

45. Ryder, *Voightlander and I,* ch. 1.

46. Marcus Root, *The Camera and the Pencil,* 26.

47. F. O. Matthiessen, *American Renaissance,* 234.

48. Ryder, *Voightlander and I,* 16.

49. Nathaniel Hawthorne, *The Blithedale Romance* (Columbus: Ohio State Univ. Press, 1970), 226. All subsequent references to this book noted by page number in the text.

50. Nathaniel Hawthorne, "Outside Glimpses of English Poverty," in *Our Old Home* (Columbus: Ohio State Univ. Press, 1970), 277.

51. *Ibid.,* 304.

52. *Ibid.,* 299.

Chapter 2. Henry James and Alvin Langdon Coburn

1. Sadakichi Hartmann, "The Broken Plates," *Camera Work* VI (April 1904): 35-39.

2. *Ibid.*

3. Henry Peach Robinson, *The Elements of a Pictorial Photograph* (1896; rpt. New York: Arno, 1973), 20.

4. *Ibid.,* 5.

5. *Ibid.*, 13.

6. Peter Henry Emerson, "Science and Art," in *Photographers on Photography,* ed. Nathan Lyons (Englewood Cliffs, N.J.: Prentice-Hall and George Eastman House, 1966), 63.

7. Beaumont Newhall, *The History of Photography* (New York: Museum of Modern Art, 1964).

8. Sadakichi Hartmann, "The Broken Plates," 37.

9. Alvin Langdon Coburn, "Henry James and the Camera," (London: B.B.C. Third Programme, 18 July 1953, 9:50–10:15 p.m.), 1.

10. Henry James to Alvin Langdon Coburn, 26 July 1906, MS, Collections 6251-9, Box 4, Alderman Library, Univ. of Virginia. Note: James misdated this letter as 26 June 1906.

11. Henry James, "The Golden Bowl," in *The Art of the Novel* (New York: Scribner's, 1934), 333.

12. *Ibid.*

13. Henry James to Alvin Langdon Coburn, 12 Aug. 1912, MS, Collection 6251-9, Box 5, Alderman Library, Univ. of Virginia.

14. Henry James, "The Real Thing," in *The Aspern Papers and Other Stories* (Harmondsworth, Middlesex, England: Penguin, 1976), 120.

15. Henry James, *The Aspern Papers,* 42.

16. Sadakichi Hartmann, "A New Departure in Photography," *The Lamp* XXVIII (Feb. 1904): 25.

17. Alvin Langdon Coburn, *Alvin Langdon Coburn Photographer: An Autobiography,* ed. Helmut and Alison Gernsheim (New York: Frederick A. Praeger, 1966), 44.

18. *Ibid.*, 46.

19. Henry James, "The Awkward Age," in *The Art of the Novel,* 101.

20. Henry James, "The Aspern Papers," in *The Art of the Novel,* 163.

21. Susan Sontag, *On Photography* (New York: Farrar, Straus and Giroux, 1978), 159.

22. George J. Becker, *Documents of Modern Literary Realism* (Princeton: Princeton Univ. Press, 1963), 165.

23. Henry James, "Balzac," in *Future of the Novel,* ed. Leon Edel (New York: Vintage, 1956), 108.

24. Henry James, "Zola," in *The Art of the Novel,* 158.

25. *Ibid.*, 168, 161.

26. Henry James, "Guy de Maupassant," in *Future of the Novel,* 195.

27. Henry James, "Balzac," in *Future of the Novel,* 122.

28. Alvin Langdon Coburn, *An Autobiography,* 44.

29. Henry James to Alvin Langdon Coburn, 7 Dec. 1906, MS, Collection 6251-9, Box 4, Alderman Library, Univ. of Virginia.

30. Alvin Langdon Coburn, *An Autobiography,* 58.

31. Henry James, "The Golden Bowl," 335.

32. Henry James to Alvin Langdon Coburn, 2 Oct. 1906, MS, Collection 6251-9, Box 4, Alderman Library, Univ. of Virginia.

33. Henry James to Alvin Langdon Coburn, 6 Dec. 1906, MS, Collection 6251-9, Box 4, Alderman Library, Univ. of Virginia.

34. *Ibid.*

35. Beaumont Newhall, *The History of Photography,* 111.

36. George J. Becker, *Documents of Modern Literary Realism,* 14.

37. Alvin Langdon Coburn, *An Autobiography*, 24.

38. Henry James, "Balzac," in *Future of the Novel*, 107.

39. Henry James, "The Golden Bowl," 327.

40. Mark Seltzer, "The Princess Casamassima: Realism and the Fantasy of Surveillance," *American Realism: New Essays*, ed. Eric J. Sundquist (Baltimore: Johns Hopkins Univ. Press, 1982), 97.

41. Henry James, *The Sacred Fount* (New York: Scribner's, 1908), 34. All subsequent references to this book noted by page number in the text.

42. Henry James, *The Princess Casamassima* (New York: Scribner's, 1908), I, p. 200. All subsequent references to this book noted by volume and page number in the text.

43. Jane Addams, et al., *Philanthropy and Social Progress* (New York: 1893), 1–26.

44. Jeffrey Mehlman, *Revolution and Repetition: Marx/Hugo/Balzac* (Berkeley & Los Angeles: Univ. of California Press, 1977).

45. Henry James, "Author's Preface," *The Princess Casamassima*, I, 20.

46. Alvin Langdon Coburn, *An Autobiography*, 80.

47. Article dated 2 Oct. 1894, in J. Fishman, *Jewish Radicals from Czarist Stetl to London Ghetto* (New York: Pantheon, 1974), 315.

48. Alan Trachtenberg, "Experiments in Another Country: Stephen Crane's City Sketches," in *American Realism: New Essays.*

49. Jack London, *The People of the Abyss* (London: Macmillan, 1903), vii. This book, like the first hardcover edition of William D. Howell's *London Films*, is copiously illustrated with photographs not attributed to any source.

50. *Ibid.*, 225.

51. *Ibid.*, 19.

52. *Ibid.*, 86.

53. William D. Howells, *London Films* (London: Harper and Bros., 1906), 106.

54. Mrs. Cecil Chesterton, *In Darkest London* (New York: Macmillan, 1926), vii.

55. General [William] Booth, *In Darkest England and the Way Out* (London: International Headquarters of the Salvation Army, 1890).

56. Gareth Stedman Jones, *Outcast London* (Oxford: Clarendon, 1971), 285.

57. *Ibid.*, 291.

58. *Ibid.*, 281.

59. Henry Mayhew, *London Labor and the London Poor* (London: Griffin, Bohn, 1861–62), Vols. I–IV; George Sims, *The Mysteries of Modern London* (London: C. Arthur Pearson, 1906); Adolphe Smith, *Street Life in London* with photographs by John Thompson (1877, rpt. Bronx, N.Y.: Benjamin Blum, 1969).

Chapter 3. Theodore Dreiser, Alfred Stieglitz, and Jacob Riis

1. Henry James, *The American Scene* (New York: Horizon, 1967), 86, 35. All subsequent references to this book noted by page number in the text.

2. Charles Loring Brace, *The Dangerous Classes of New York* (New York: Wynkoop and Hallenbeck, 1872), 29.

3. Henry Adams, *The Education of Henry Adams* (Boston: Houghton Mifflin, 1974), 120.

4. Theodore Dreiser, "Reflections," *Ev'ry Month* III (Oct. 1896): 6–7.

5. Alfred Stieglitz, "Pictorial Photography," in *Classic Essays on Photography*, ed. Alan Trachtenberg (New Haven: Leete's Island Books, 1980), 122.

6. Jacob Riis, *How the Other Half Lives* (1896; rpt. New York: Dover, 1971), 209.

7. Theodore Dreiser, "A Master of Photography," *Success* (10 June 1899): 471.

8. Theodore Dreiser, "A Remarkable Art: The New Pictorial Photography," *The Great Round World* (3 May 1902): 430.

9. *Ibid.*, 433, 434.

10. Theodore Dreiser, *Newspaper Days* (New York: Horace Liveright, 1922), 100.

11. Theodore Dreiser, *A Book about Myself* (New York: Boni and Liveright, 1922), 463–64.

12. Theodore Dreiser, *Newspaper Days,* 449.

13. *Ibid.*, 454.

14. Mary Antin, *The Promised Land* (Boston: Houghton Mifflin, 1912), 88.

15. Paul Rosenfeld, "Alfred Stieglitz," *The Dial* LXX (April 1921): 409.

16. Robert Haines, "Alfred Stieglitz and the New Order of Consciousness in American Literature," *Pacific Coast Philology* VI (April 1971): 28.

17. Paul Rosenfeld, "Alfred Stieglitz," *Port of New York* (New York: Harcourt, Brace, 1924), 257.

18. Dorothy Norman, ed., "From the Writings and Conversations of Alfred Stieglitz," *Twice-a-Year* I (Fall–Winter 1938): 110.

19. Hart Crane, "The Tunnel," and "Atlantis," *The Bridge* (New York: Liveright, 1970), 69, 74.

20. Marius de Zayas in *291,* 5–6 (July–Aug. 1915).

21. Alfred Stieglitz, "How I Came to Photograph Clouds," *The Amateur Photographer and Photography* LVI (1923): 255.

22. Dorothy Norman, *Alfred Stieglitz: An American Seer* (Millerton, N. Y.: Aperture, 1973), 161.

23. *Ibid.*

24. Charles Caffin, *Camera Works* XXV (Jan. 1909): 17.

25. Marius de Zayas, *Camera Works* XXXIV–XXXV (April–July 1911): 66.

26. Dorothy Norman, *Alfred Stieglitz: An American Seer,* 161.

27. Herbert Seligmann, *Alfred Stieglitz Talking* (New Haven: Yale Univ. Library, 1966), ?.

28. Dorothy Norman, *Alfred Stieglitz: An American Seer,* 28.

29. *Ibid.*, 9.

30. John Carlos Rowe, *The Theoretical Dimensions of Henry James* (Madison: Univ. of Wisconsin Press, 1984).

31. "Alfred Stieglitz and His Latest Work," *The Photographic Times* XXVIII (April 1896): 161.

32. Dorothy Norman, *Alfred Stieglitz: An American Seer,* 9.

33. "Alfred Stieglitz: Four Happenings," *Twice-a-Year* VIII–IX (1942): 130.

34. *Ibid.*, 132.

35. Elizabeth Bowen, *Death of the Heart* (New York: Knopf, 1939), 45.

36. Edith Wharton, *The Age of Innocence* (New York: Scribner's, 1968), 108.

37. Paul Rosenfeld, *Port of New York,* 238.

38. Waldo Frank, "The New World in Alfred Stieglitz," *American and Alfred Stieglitz* (Millerton, N.Y.: Aperture, 1979), 109.

39. Stuart P. Sherman, "The Naturalism of Mr. Dreiser," in *Documents of Modern Literary Realism,* ed. George J. Becker (Princeton: Princeton Univ. Press, 1963), 453, 455.

40. Theodore Dreiser, *Newspaper Days,* 491.

41. Paul Rosenfeld, *Port of New York,* 237.

42. Theodore Dreiser, "A Remarkable Art," 430.

43. Theodore Dreiser, *Theodore Dreiser: A Selection of Uncollected Prose*, ed. Donald Pizer (Detroit: Wayne State Univ. Press, 1977), 177.

44. Theodore Dreiser, *The Color of a Great City* (New York: Boni and Liveright, 1923), 68.

45. *Ibid.*, 206.

46. Theodore Dreiser, *An Amateur Laborer*, ed. Richard W. Dowell (Philadelphia: Univ. of Pennsylvania Press, 1983), 164-65.

47. *Ibid.*, 52.

48. *Ibid.*, 25.

49. *Ibid.*, 56.

50. Theodore Dreiser, *The Color of a Great City*, 77.

51. Theodore Dreiser, *An Amateur Laborer*, 161.

52. Theodore Dreiser, *Hey Rub-a-Dub-Dub: A Book of the Mystery and Wonder and Terror of Life* (New York: Boni and Liveright, 1920), 94.

53. Theodore Dreiser, *An Amateur Laborer*, 172.

54. Theodore Dreiser, *The Color of a Great City*, 129.

55. Theodore Dreiser, "The Literary Shower," *Ev'ry Month* I (Feb. 1896): 47.

56. Theodore Dreiser, *An Amateur Laborer*, xxxvii.

57. Theodore Dreiser, *The Color of a Great City*, 169.

58. John Berger, "The Primitive and the Professional," *About Looking* (New York: Pantheon, 1980), 68.

59. Charles Loring Brace, *Dangerous Classes*, 282.

60. Jacob Riis, *The Making of an American* (New York: Macmillan, 1904), 316.

61. Mary O. Furner, *Advocacy and Objectivity: A Crisis in the Professionalization of American Social Science 1865-1905* (Lexington: Univ. of Kentucky Press, 1975), 31-32.

62. Edward L. Youmans, "Under False Colors," *Popular Science Monthly* VII (July 1875): 365-66.

63. See Brian Fay, *Social Theory and Political Practice* (London: Allen and Unwin, 1975).

64. Charles Loring Brace, *Dangerous Classes*, 29.

65. Jacob Riis, *How the Other Half Lives*, 209.

66. *Ibid.*, 133.

67. *Ibid.*, 90.

68. *Ibid.*, 52.

69. Jacob Riis, *The Making of an American*, 66-67.

70. Jacob Riis, *How the Other Half Lives*, 54.

71. *Ibid.*, 174.

72. *Ibid.*, 88.

73. Jacob Riis, *The Making of an American*, 267.

74. *Ibid.*, 268.

75. *Ibid.*, 59.

76. Peter B. Hales, *Silver Cities: The Photography of American Urbanization, 1839-1915* (Philadelphia: Temple Univ. Press, 1984), 87.

77. Charles Loring Brace, *Dangerous Classes*, 119.

78. Theodor Adorno, *Negative Dialectics*, trans. E. B. Ashton (New York: Seabury, 1973).

79. Theodore Dreiser, *The "Genius"* (New York: John Lane, 1915), 223. All subsequent references to this book noted by page number in the text.

80. Theodore Dreiser, *Newspaper Days*, 183.

81. See correspondence between Dreiser and Joseph Coates: Coates to T. D., 18 July 1912, 20 Aug. 1912, Box 52, Theodore Dreiser Collection, Van Pelt Library, Univ. of Pennsylvania.

82. Theodore Dreiser, *The Financier* (New York: Harper and Bros., 1912), 12.

83. *Ibid.*, 313.

Chapter 4. John Dos Passos and the Soviet Cinema

1. John Dos Passos, *Manhattan Transfer* (1925; rpt. Cambridge, Mass.: Robert Bentley, 1980), 365, 194.

2. Jeremiah 38:2.

3. John Dos Passos, *Manhattan Transfer*, 320, 344.

4. John Dos Passos to Rumsey Marvin, 1916, in Townsend Ludington, ed., *The Fourteenth Chronicle: Letters and Diaries of John Dos Passos* (Boston: Gambit, 1973), 39. See also Townsend Ludington, *John Dos Passos: A Twentieth Century Odyssey* (New York: E. P. Dutton, 1980), 81.

5. John Dos Passos to Dudley Poore, 15 Sept. 1916, in Townsend Ludington, *John Dos Passos*, 90.

6. *Ibid.*, 218.

7. John Dos Passos, *Manhattan Transfer*, 360.

8. John Dos Passos, *Century's Ebb*, quoted in *The Fourteenth Chronicle*, 398.

9. John Dos Passos, *Manhattan Transfer*, 375.

10. *Ibid.*, 386, 390.

11. Townsend Ludington, *John Dos Passos*, 237.

12. *Ibid.*, 269, 270.

13. Seth R. Feldman, *Dziga Vertov: A Guide to References and Resources* (Boston: G. K. Hall, 1979) 11. See also Peter Dart, *Pudovkin's Films and Film Theory* (New York: Arno, 1974).

14. Seth Feldman, *Dziga Vertov*, 11. For further discussion of Dos Passos and Vertov in the Soviet Union, see Vlada Petric, *Constructivism in Film: "The Man with the Movie Camera." Cinematic Analysis* (Cambridge: Cambridge Univ. Press, 1987).

15. Seth Feldman, *Dziga Vertov*, 25, 26.

16. P. A. Sitney, ed., "The Writings of Dziga Vertov," in *Film Culture Reader* (New York: Praeger, 1970), 362.

17. *Ibid.*, 373–74.

18. Seth Feldman, *Dziga Vertov*, 4.

19. P. A. Sitney, "The Writings of Dziga Vertov," 356–57.

20. Lázló Maholy-Nagy, "From Pigment to Light," in *Photographers on Photography*, ed. Nathan Lyons (Englewood Cliffs, N.J.: Prentice-Hall, 1966), 80.

21. Erik Barnouw, *Documentary: A History of the Non-fiction Film* (New York: Oxford Univ. Press, 1974), 60; see also Jay Leyda, *Kino: A History of the Russian and Soviet Film* (New York: Collier, 1973); see also J. Mayne, *The Ideologies of Metacinema* (Ann Arbor, Mich.: University Microfilms, 1975).

22. Alfred Kazin, "Introduction," *The Big Money*, by John Dos Passos (Boston: Houghton Mifflin, 1969), x.

23. *Ibid.*, ix.

24. *Ibid.*

25. Townsend Ludington, *John Dos Passos, 259.*

26. *Ibid.,* 260.

27. P. A. Sitney, "The Writings of Dziga Vertov," 374.

28. *Ibid.,* 359.

29. John Dos Passos, *The 42nd Parallel* (Boston: Houghton Mifflin, 1946), 104.

30. Townsend Ludington, *John Dos Passos,* 257.

31. John Dos Passos, *The Big Money,* 523.

32. *Ibid.,* 444.

33. *Ibid.,* 469.

34. John Dos Passos, "The Writer as Technician," in *American Writers' Congress,* ed. Henry Hart (New York: International Publishers, 1935), 82.

35. John Dos Passos, *Journeys between Wars* (New York: Harcourt, Brace, 1930), 237.

36. John Dos Passos, "The Writer as Technician," 79, 81.

37. John Dos Passos, *The Theme Is Freedom* (New York: Dodd, Mead, 1956), 115.

38. See H. Solow, "Substitution at Left Tackle: Hemingway for Dos Passos," *Partisan Review* IV, No. 5 (April 1938).

39. Joris Ivens, *The Camera and I* (New York: International Publishers, 1969), 52; see also Wolfgang Klaue, Hans Wegner, and Manfred Lichtenstein, *Joris Ivens: Zusammenstellung und Redaktion* (Berlin: Staatlichen Filmarchiv der Republik, 1963); see also Hans Wegner, *Joris Ivens: Dokumentarist der Warhrheit* (Berlin: Henschelverlag Kunst und Gesellschaft, 1965).

40. *Ibid.,* 59.

41. *Ibid.,* 88.

42. *Ibid.,* 91.

43. *Ibid.,* 75.

44. *Ibid.,* 77.

45. John Dos Passos, "Spain: Rehearsal for Defeat," *The Theme Is Freedom,* 116.

46. William B. Watson and Barton Whaley, *The Spanish Earth of Dos Passos and Hemingway,* TS. I wish to thank Elinor Langer for access to this material.

47. Joris Ivens, *The Camera and I,* 110.

48. *Ibid.,* 111.

49. William B. Watson and Barton Whaley, *The Spainish Earth of Dos Passos and Hemingway.*

50. *Ibid.,*

51. Joris Ivens, *The Camera and I,* 107.

52. *Ibid.,* 108–9.

53. *Ibid.,* 111.

54. *Ibid.,* 120.

55. William Watson, and Barton Whaley, *The Spanish Earth,* 18.

56. *Ibid.,* 12.

57. *Ibid.*

58. P. A. Sitney, "The Writings of Dziga Vertov," 359, 362.

59. Joris Ivens, *The Camera and I,* 103.

60. William Watson, and Barton Whaley, *The Spanish Earth,* 46.

61. *Ibid.,* 44.

62. John Dos Passos, *Journeys between Wars,* 392–93.

63. *Ibid.,* 394.

Chapter 5. James Agee and Walker Evans

1. Walker Evans, "James Agee in 1936," in *Let Us Now Praise Famous Men* (Cambridge: Houghton Mifflin, 1960), xi, xii.

2. *Ibid.,* xi.

3. James Agee, *Let Us Now Praise Famous Men* (Cambridge: Houghton Mifflin, 1960), 7. All subsequent references to this book noted by page number in the text.

4. Margaret Bourke-White, *The Photographs of Margaret Bourke-White,* ed. Sean Callahan (Boston: New York Graphic Society, 1972), 13.

5. In William Stott, *Documentary Expression and Thirties America* (New York: Oxford Univ. Press, 1973), 216.

6. *Ibid.,* 222.

7. Margaret Bourke-White, *You Have Seen Their Faces* (New York: Viking, 1937), 187.

8. *Ibid.,* 189.

9. Margaret Bourke-White, *Portrait of Myself* (New York: Simon and Schuster, 1963), 125, 126–27.

10. Margaret Bourke-White, *You Have Seen Their Faces,* 187.

11. Mildred Gwin Barnwell, *Faces We See* (Gastonia, N.C.: Southern Combed Yarn Spinners Assoc., 1939), 17.

12. *Ibid.,* 18.

13. William Stott, *Documentary Expression,* 223.

14. Edward Weston, *The Daybooks of Edward Weston* (Rochester, N.Y.: George Eastman House, n.d.), I, 72.

15. William Stott, "Walker Evans: A Memoir and Introduction," TS, quoted by permission of the author.

16. Beaumont Newhall, *The History of Photography* (New York: Museum of Modern Art, 1964), 113.

17. *Ibid.,* 114.

18. William Stott, "Walker Evans," 16–17.

19. *Ibid.,* 7.

20. In Edward Weston, *Daybooks,* I, 5.

21. Henry James, "Preface to *The Golden Bowl,*" in *The Art of the Novel* (New York: Scribner's 1937), 327.

Chapter 6. John Steinbeck and Dorothea Lange

1. "Interview with Richard K. Doud for the Archives of American Art as Part of Its New Deal and the Arts Project," TS, 22 May 1964, Fogg Art Museum Library, Harvard University.

2. Carl Mydans, *More than Meets the Eye* (New York: Harper and Bros., 1959), 310.

3. "Interview with Richard Doud," 67.

4. See Milton Meltzer, *Dorothea Lange: A Photographer's Life* (New York: Farrar, Straus, and Giroux, 1978); see also Robert Coles, "Essay," in *Dorothea Lange: Photographs of a Lifetime* (Millerton, N.Y.: Aperture, 1982).

5. John Berger, "Paul Strand," in *About Looking* (New York: Pantheon, 1980), 43.

6. "Dorothea Lange: The Making of a Documentary Photographer," Interview by Suzanne Riess for the Regional Oral History Office, TS, 1968, Fogg Art Museum Library, Harvard University.

7. *Ibid.;* see also Dorothea Lange and Paul Schuster Taylor, *An American Exodus: A Record of Human Erosion* (New York: 1939; rpt. New Haven: Yale Univ. Press in association with the Oakland Museum, 1969).

8. *Ibid.,* 149.

9. *Ibid.,* 161.

10. Michel Foucault, "The Eye of Power," in *Power/Knowledge* (New York: Pantheon, 1980), 152.

11. Paul Schuster Taylor, "Establishment of Rural Rehabilitation Camps for Migrants in California," Emergency Relief Administration, 15 March 1935, in *Dorothea Lange Farm Security Administration Photographs 1935-1939,* ed. Howard M. Levin and Katherine Northrup (Glencoe, Ill.: Text-Fiche Press, 1980), II, p. 77.

12. *Ibid.,* 66.

13. In Suzanne Riess, "Dorothea Lange," 175.

14. Karin B. Ohrn, *Dorothea Lange and the Documentary Tradition* (Baton Rouge: Louisiana State Univ. Press, 1980), 56.

15. In "Interview with Richard Doud," 68.

16. *Ibid.*

17. *Ibid.,* 57.

18. Michel Foucault, "The Eye of Power," in *Power/Knowledge,* 154.

19. Karin Ohrn, *Dorothea Lange,* 57.

20. *Ibid,.* 56.

21. In "Interview with Richard Doud," 65-68.

22. Karin Ohrn, *Dorothea Lange,* 233.

23. *Ibid.,* 146, 133.

24. D. G. Kehl, "Steinbeck's 'String of Pictures' in *The Grapes of Wrath,*" *Image* XVII, No. 1 (March 1974): 4.

25. *Ibid.,* 2.

26. See the John Steinbeck Papers, Humanities Research Center, University of Texas at Austin.

27. Jackson J. Benson, "'To Tom, Who Lived It': John Steinbeck and the Man from Weedpatch," *Journal of Modern Literature* (1976): 174; see also Jackson J. Benson, *The True Adventures of John Steinbeck, Writer* (New York: Viking, 1984).

28. John Steinbeck, "Their Blood Is Strong," in *A Companion to the Grapes of Wrath,* ed. Warren French (New York: Viking, 1963), 54-65.

29. *Ibid.,* 56.

30. The John Steinbeck Papers.

31. John Steinbeck, "Their Blood Is Strong," 63.

32. In Jackson Benson, "The Man from Weedpatch," 181-82.

33. *Ibid.,* 183.

34. In Jackson Benson, *True Adventures,* 369.

35. John Steinbeck, *In Dubious Battle* (New York: Covici-Friede, 1936), 140. All subsequent references to this book noted by page number in the text.

36. See Elizabeth Janeway, *Powers of the Weak* (New York: William Morrow, 1981).

37. John Steinbeck, *The Grapes of Wrath* (New York: Viking, 1939), 64.

38. See Lewis Hyde, *The Gift: Imagination and the Erotic Life of Property* (New York: Vintage, 1983); see also Erving Goffman, *Interaction Ritual* (New York: Pantheon, 1967).

39. Jackson Benson, "The Man from Weedpatch," 173.

40. *Ibid.*, 213.

41. Tom Collins, "Bringing in the Sheaves," *Journal of Modern Literature* II (1976): 226.

42. See Martin Staples Shockley, "The Reception of *The Grapes of Wrath* in Oklahoma," *American Literature* XV (May 1944): 351-61.

43. Tom Collins, "Bringing in the Sheaves," 228, 230.

44. *Ibid.*, 232.

45. John Steinbeck, *Steinbeck: A Life in Letters,* ed. Elaine Steinbeck and Robert Wallsten (New York: Viking, 1975), 119.

46. *Ibid.*

Chapter 7. Norman Mailer and Combat Photography

1. Norman Mailer, "First Advertisement for Myself," in *The Long Patrol: Twenty-five Years of Writing from the World of Norman Mailer,* ed. Robert F. Lucid (New York: World Publishers, 1971), 158.

2. "Norman Mailer," *Current Biography* (Oct. 1948): 409. For more information about the 112th Cavalry see Major B. C. Wright, *The First Division in World War II* (Tokyo: Toppan Printing, 1947).

3. Norman Mailer, *The Naked and the Dead* (New York: Rinehart, 1948), 84. All subsequent references to this book noted by page number in the text.

4. Col. Roy M. Stanley II, *World War II Photo Intelligence* (New York: Scribner's, 1981), 65, 66; see also Norman Barr Moyes, *Major Photographers and the Development of Still Photography in Major American Wars* (Ann Arbor, Mich.: University Microfilms, 1967).

5. George Goddard, *Overview: A Life long Adventure in Aerial Photography* (Garden City, N.Y.: Doubleday, 1969), 334.

6. U.S. Army Technical Manual, TM 11-401-2: Army Pictorial Techniques, Equipments and Systems, Still Photography, 12-1.

7. Roy Stanley, *Photo Intelligence,* 247; see also U.S. Army Technical Manual, TM 30-245: Image Interpretation Handbook; see also Defense Intelligence Agency Publications, DIAM 55-5: Aerial Photography and Airborne Electronic Sensor Imagery.

8. Margaret Bourke-White, *Purple Heart Valley: A Chronicle of the War in Italy* (New York: Simon and Schuster, 1944), 5; see also Margaret Bourke-White, *Shooting the Russian War* (New York: Simon and Schuster, 1942).

9. Roy Stanley, *Photo Intelligence,* 68.

10. Michel Foucault, "Powers and Strategies," in *Power/Knowledge* (New York: Pantheon, 1980), 142.

11. G. R. Thompson and Dixie R. Harris, *The Technical Services: The Signal Corps: The Outcome* (Washington, D.C.: Office of the Chief of Military History, United States Army, 1966), 540.

12. Robert Eichberg and Jacqueline Quadow, Signal Corp Historical Project F-2b: *Combat Photography,* Nov. 1945, TS, Modern Military Branch, Military Archives Division, National Archives and Records Service, Washington, D.C.: x-xi, 54.

13. Carl Mydans, *More than Meets the Eye* (New York: Harper and Bros., 1959), 4, 8.

14. Letter of Col. M. E. Gillette, A.P.S. Signal Section, H.Q., Fifth Army to Deputy CSigO A.F.H.Q., 16 Jan. 1944 at Department of the Army: Center for Military

History, Washington, D.C.; see also "Exposure under Fire: An Official History of SigC Photography in the Luzon Operation," U.S. Army SigC, S.W.P.A., 25 April 1945.

15. Robert Capa, *Images of War* (New York: Paragraphic Books, 1964), 62, 77, 84; see also Edward R. Trabold, "Counterattacking with a Camera," *Photographic Journal of America* CVI (1919): 407–11; and Jerry Joswick and Lawrence Keating, *Combat Cameraman* (Philadelphia: Chilton, 1961).

16. John Berger, "Uses of Photography," in *About Looking* (New York: Pantheon, 1980), 58.

17. Robert Capa, *Images of War*, 82.

18. "Norman Mailer," *Current Biography*, 409.

19. Jan Valtin [pseud.], *Children of Yesterday* (New York: Readers' Press, 1946), 9.

20. See M. Hamlin Cannon, *The War in the Pacific: Leyte: The Return to the Philippines*, United States Army in World War II Series (Washington, D.C.: Office of the Chief of Military History, United States Army, 1954); see also Robert Ross Smith, *The War in the Pacific: Triumph in the Philippines*, United States Army in World War II Series (Washington, D.C.: Office of the Chief of Military History, United States Army, 1963).

21. See Stanley L. Falk, *Decision at Leyte* (New York: W. W. Norton, 1966); also Peyton Hodge, Thomas Hooper, and Victor Lott, *Avengers of Bataan, 38th Infantry Division, Luzon Campaign* (Atlanta: Albert Lore Enterprises, 1947); also Harold Whittle Blakeley, *The 32nd Infantry Division in World War II* (Madison: 1957); also *13,000 Hours: Combat History of the 32nd Infantry Division World War II*, prepared by the Public Relations Office, 32nd Inf. Div. under the direction of AC of S, G-Z (Manila: 1945).

22. Robert Capa, *Images of War*, 74.

23. Jan Valtin, *Children of Yesterday*.

24. Norman Mailer, "The White Negro," in *The Long Patrol*, 225.

25. Norman Mailer, "Some Dirt in the Talk: A Candid History of an Existential Movie Called *Wild 90*," *Esquire* LXVIII (Dec. 1967): 269.

26. *Ibid.*, 190, 269.

27. Robert Capa, *Images of War*, 62.

28. Pauline Kael, "Celebrities Make Spectacles of Themselves," *New Yorker* XX (Jan. 1968): 90.

29. Norman Mailer, "The White Negro," 215.

30. Norman Mailer, *The Deer Park* (New York: Putnam Publishing Group, 1981), 68.

31. Norman Mailer, "The White Negro," 210, 211.

32. Norman Mailer, "*Wild 90*," 194.

33. *Ibid.*, 192.

34. Fredric Jameson, *The Political Unconscious* (Ithaca: Cornell Univ. Press, 1981), 299.

Conclusion

1. Michel Foucault, *Discipline and Punish*, trans. Alan Sheridan (New York: Vintage, 1979), 202.

2. René Girard, *Deceit, Desire and the Novel*, trans. Yvonee Freccero (Baltimore: Johns Hopkins Univ. Press, 1965), 300.

3. See Robert Weimann, *Structure and Society in Literary History* (Charlottesville: Univ. Press of Virginia, 1976), 237. "[The writer faces] a world full of struggle and

change where the writer, in order to transmute his experience into art, has constantly to reassess his relations to society as both a social and an aesthetic act. In the process of doing this, he will find that his own experience as an artist in history is so related to the social whole that the flexibility (which involves the precariousness) of this relationship itself is the basis on which representation and evaluation are integrated through point of view." See also Susan Sniader Lanser, *The Narrative Act: Point of View in Prose Fiction* (Princeton: Princeton Univ. Press, 1981).

4. Philip Caputo, *DelCorso's Gallery* (New York: Holt, Rinehart and Winston, 1983), 224.

5. *Ibid.*

6. *Ibid.,* 321.

7. Paul Theroux, *Picture Palace* (New York: Ballantine, 1978), 243.

8. *Ibid.,* 236.

9. Fredric Jameson, *The Political Unconscious* (Ithaca, N.Y.: Cornell Univ. Press, 1981), 299.

10. Lucas Samaras, *Samaras Album* (New York: Whitney Museum of Art, 1971), 5.

11. *Ibid.,* 3.

12. *Ibid.*

13. Peter Schjeldahl, "Introduction," to Cindy Sherman, *Cindy Sherman* (New York: Pantheon, 1984).

14. Eudora Welty, *One Writer's Beginnings* (Cambridge: Harvard Univ. Press, 1984), 85.

15. *Ibid.,* 84.

16. *Ibid.,* 87.

17. *Ibid.,* 107.

18. *Ibid.*

19. John Berger, *Another Way of Telling* (New York: Pantheon, 1982), 279.

20. *Ibid.*

21. *Ibid.,* 287.

22. Richard Rudisill, *Mirror Image: The Influence of the Daguerreotype on American Society* (Albuquerque: Univ. of New Mexico Press, 1971), 48.

23. M. M. Bakhtin, *The Dialogic Imagination,* trans. Caryl Emerson and Michael Holquist (Austin: Univ. of Texas Press, 1981), 252–53.

BIBLIOGRAPHY

Aaron, Daniel, and Robert Bendiner. *The Strenuous Decade: A Social and Intellectual Record of the 1930s.* Garden City, N.Y.: Doubleday, 1970.

Adams, Henry. *The Education of Henry Adams.* Boston: Houghton Mifflin, 1974.

Addams, Jane, *et al. Philanthropy and Social Progress.* New York: 1893.

Adorno, Theodor. *Negative Dialectics.* Trans. E. B. Ashton, New York: Seabury, 1973.

Agee, James. *Agee on Film: Reviews and Comments.* I. New York: McDowell, Obolensky, 1958.

———. "Art for What's Sake?" *New Masses* XXI (15 Dec. 1936).

———. *The Collected Short Prose of James Agee.* Ed. Robert Fitzgerald. Boston: Houghton Mifflin, 1962.

———. *The Letters of James Agee to Father Flye.* New York: George Braziller, 1962.

———. *Let Us Now Praise Famous Men.* 1941; rpt. Boston: Houghton Mifflin, 1960.

———. "Sharecropper Novels." *New Masses* XXI (8 June 1937).

———. *A Way of Seeing, Photographs of New York by Helen Levitt.* New York: Viking, 1965.

Alexander, William. *Film on the Left: American Documentary Film from 1931 to 1942.* Princeton: Princeton Univ. Press, 1981.

Alland, Alexander. *Jacob A. Riis: Photographer and Citizen.* New York: Aperture, 1974.

Anderson, Sherwood. *Hometown.* With photographs by the Farm Security Administration. New York: Alliance Book Corp., 1940.

Antin, Mary. *The Promised Land.* Boston: Houghton Mifflin, 1912.

Arbus, Diane. *Diane Arbus.* Millerton, N.Y.: Aperture, 1972.

Arnheim, Rudolph. "On the Nature of Photography." *Critical Inquiry* (Autumn 1974).

Arthur, T. S. "The Daguerreotypist." *Godey's Lady's Book* (May 1849).

Bakhtin, M. M. *The Dialogic Imagination.* Trans. Caryl Emerson and Michael Holquist. Austin: Univ. of Texas Press, 1981.

Baldwin, Sidney. *Poverty and Politics: The Rise and Decline of the Farm Security Administration.* Chapel Hill, N.C.: Univ. of North Carolina Press, 1968.

Barnouw, Erik. *Documentary: A History of the Non-fiction Film.* New York: Oxford Univ. Press, 1974.

Barnwell, Mildred Gwin. *Faces We See.* Gastonia, N.C.: Southern Combed Yarn Spinners Assoc., 1939.

Barson, Alfred T. *A Way of Seeing: A Critical Study of James Agee.* Amherst: Univ. of Massachusetts Press, 1972.

Barthes, Roland. *Camera Lucida: Reflections on Photography.* Trans. Richard Howard. New York: Hill and Wang, 1981.

———. *Mythologies.* Trans. Annette Lavers. New York: Hill and Wang, 1972.

Baudelaire, Charles. "Le public moderne et la photographie." *Mirror of Art.* 1859; rpt. London: Phaidon, 1955.

Bazin, André. "Ontology of the Photographic Image." In *What Is Cinema?* 2 vols. Trans. Hugh Gray. Berkeley: Univ. of California Press, 1967.

Becker, George J., ed. *Documents of Modern Literary Realism.* Princeton: Princeton Univ. Press, 1963.

Benson, Jackson J. "'To Tom, Who Lived It': John Steinbeck and the Man from Weedpatch." *Journal of Modern Literature* (1976).

————. *The True Adventures of John Steinbeck, Writer.* New York: Viking, 1984.

Berger, John. *About Looking.* New York: Pantheon, 1980.

————. *Another Way of Telling.* New York: Pantheon, 1982.

————. *Pig Earth.* New York: Pantheon, 1979.

————. *Ways of Seeing.* London: Penguin, 1972.

Bergreen, Laurence. *James Agee: A Life.* New York: E. P. Dutton, 1984.

Berger, Peter L., and Thomas Luckmann. *The Social Construction of Reality.* New York: Anchor, 1967.

Bisztray, George. *Marxist Models of Literary Realism* New York: Columbia Univ. Press, 1978.

Blakeley, Harold Whittle. *The 32nd Infantry in World War II.* Madison: 1957.

Block, H. M. *Naturalistic Triptych: The Fictive and the Real in Zola, Mann and Dreiser.* New York: Random, 1970.

Booth, General [William]. *In Darkest England and the Way Out.* London: International Headquarters of the Salvation Army, 1890.

Bourke-White, Margaret. *The Photographs of Margaret Bourke-White.* Ed. Sean Callahan. Boston: New York Graphic Society, 1972.

————. *Portrait of Myself.* New York: Simon and Schuster, 1963.

————. *Purple Heart Valley: A Combat Chronicle of the War in Italy.* New York: Simon and Schuster, 1944.

————. *Shooting the Russian War.* New York: Simon and Schuster, 1942.

————, and Erskine Caldwell. *You Have Seen Their Faces.* New York: Viking, 1937.

Bowen, Elizabeth. *Death of the Heart.* New York: Knopf, 1939.

Brace, Charles Loring. *The Dangerous Classes of New York.* New York: Wynkoop & Hallenbeck, 1872.

Breit, Harvey. "Cotton Tenantry." *New Republic* CV (15 Sept. 1941).

Buckland, Gail. *Reality Recorded: Early Documentary Photography.* Greenwich, Conn.: New York Graphic Society, 1974.

Buck-Morss, Susan. *The Origin of Negative Dialectics: Theodor W. Adorno, Walter Benjamin and the Frankfort Institute.* New York: Free Press, 1977.

Caffin, Charles H. *Photography as a Fine Art: The Achievements and Possibilities of Photographic Art in America.* New York: Doubleday, 1901.

Cannon, M. Hamlin. *The War in the Pacific: Leyte: The Return to the Philippines.* United States Army in World War II Series. Washington, D.C.: Office of the Chief of Military History, United States Army, 1954.

Cantor, Jay. *The Space Between: Literature and Politics.* Baltimore: Johns Hopkins Univ. Press, 1981.

Capa, Robert. *Images of War.* New York: Paragraphic Books, 1964.

————. *Slightly Out of Focus.* New York: Henry Holt, 1947.

Caputo, Philip. *DelCorso's Gallery.* New York: Holt, Rinehart and Winston, 1983.

Cartier-Bresson, Henri. "The Decisive Moment." In *Photographers on Photography.* Ed. Nathan Lyons. Englewood Cliffs, N.J.: Prentice-Hall, 1966.

Carver, Craig. "The Newspaper and Other Sources of *Manhattan Transfer.*" *Studies in American Fiction* III (Autumn 1975): 167–80.

Cavell, Stanley. *The World Viewed: Reflections on the Ontology of Film.* New York: Viking, 1973.

Chesterton, Mrs. Cecil. *In Darkest London.* New York: Macmillan, 1926.

Coburn, Alvin Langdon. *Alvin Langdon Coburn Photographer: An Autobiography*. Ed. Helmut and Alison Gernsheim. New York: Frederick A. Praeger, 1966.
——. "Bernard Shaw, Photographer." *Photoguide Magazine* I (Dec. 1950).
——. *Henry James and the Camera*. London: B.B.C. Third Programme, 18 July 1953, 9:50–10:15 p.m.
——. *London*. New York: Brentano, 1909.
——. *Men of Mark*. New York: Michell Kennerley, 1913.
——. *More Men of Mark*. London: Duckworth, 1922.
——. "Photography and the Quest of Beauty." *Photographic Journal* LXIIII (April 1924).
——. *A Portfolio of Sixteen Photographers*. Ed. Nancy Newhall. Rochester, N.Y.: George Eastman, 1962.
"Alvin Langdon Coburn: Artist by Himself." *Wilson's Photographic Magazine* LI (Jan. 1914).
Coles, Robert, ed. *Dorothea Lange: Photographs of a Lifetime*. Millerton, N.Y.: Aperture, 1982.
Collins, Tom. "Bringing in the Sheaves." *Journal of Modern Literature* (1976): 226–32.
Cordasco, Francesco, ed. *Jacob Riis Revisited: Poverty and the Slum in Another Era*. Clifton: Augustus M. Kelley, 1973.
Crane, Hart. *The Bridge*. New York: Liveright, 1970.
Crozier, A. T. K. "American Photography." *Journal of American Studies* XIIII (Dec. 1980): 461–65.
Cummings, Thomas H. *Photography: Its Recognition as a Fine Art and a Means of Individual Expression*. Boston: Photographers' Assoc. of America, 1905.
——— "Some Photographs by Alvin Langdon Coburn." *Photo Era* X (March 1903).
Dart, Peter. *Pudovkin's Film and Film Theory*. New York: Arno, 1974.
Dijkstra, Bram. *Cubism, Stieglitz and the Early Poetry of William Carlos Williams*. Princeton: Princeton Univ. Press, 1969.
Dos Passos, John. *The Big Money*. Boston: Houghton Mifflin, 1946.
——. *Century's Ebb: The Thirteenth Chronicle*. Boston: Gambit, 1975.
——. "Farewell to Europe." *Common Sense* VI (July 1957).
——. *The Fourteenth Chronicle: Letters and Diaries of John Dos Passos*. Ed. Townsend Ludington. Boston: Gambit, 1973.
——. *Journeys between Wars*. New York: Harcourt, Brace, 1930.
——. *Manhattan Transfer*. New York, 1925; rpt. Cambridge, Mass.: Robert Bentley, 1980.
——. *The Theme Is Freedom* New York: Dodd, Mead, 1956.
——. "The Villages Are the Heart of Spain." *Esquire* (1937).
——. "The Writer as Technician." In *American Writers' Congress*. Ed. Henry Hart. New York: International Publishers, 1935.
Doud, Richard K. "Interview with Dorothea Lange for the Archives of American Art." TS. (22 May 1964).
Draper, John W. *Scientific Memoirs*. New York: Harper and Bros., 1978.
Dreiser, Theodore, *An Amateur Laborer*. Ed. Richard W. Dowell. Philadelphia: Univ. of Pennsylvania Press, 1983.
——. *American Diaries: 1902–1926*. Ed. T. P. Riggio, et al. Philadelphia: Univ. of Pennsylvania Press, 1982.
——. "The Realist and His Sources." TS (17 April 1935). Box 172, Theodore Dreiser Collection. Van Pelt Library, Univ. of Pennsylvania.

————. *A Book about Myself.* New York: Boni & Liveright, 1922.

————. "The Camera Club of New York." *Ainslee's* (Sept. 1899).

————. *The Color of a Great City.* New York: Boni & Liveright, 1923.

————. *The Financier.* New York: Harper and Bros., 1912.

————. *The "Genius".* New York: John Lane, 1915.

————. *Hey Rub-a-Dub-Dub: A Book of the Mystery and Wonder and Terror of Life.* New York: Boni & Liveright, 1920.

————. "Life, Art and America." *The Seven Arts* (Feb. 1917).

————. "The Literary Shower." *Ev'ry Month* I (Feb. 1896).

————. "A Master of Photography." *Success* (June 1899).

————. *Newspaper Days.* New York: Horace Liveright, 1922.

————. *Notes on Life.* Ed. M. Tjader and J. J. McAleer. Birmingham: Univ. of Alabama Press, 1974.

————. "Reflections." *Ev'ry Month* III (Oct. 1896).

————. "A Remarkable Art: The New Pictorial Photography." *Great Round World* (3 May 1902).

————. *The Stoic.* New York: New American Library, 1947.

————. *Theodore Dreiser: A Selection of Uncollected Prose.* Ed. Donald Pizer. Detroit: Wayne State Univ. Press, 1977.

————. *The Titan.* New York: John Lane, 1914.

————. *Tragic America.* New York: Horace Liveright, 1931.

Eagleton, Terry. *Criticism and Ideology.* London: Verso Editions, 1978.

Edel, Leon, ed. *Henry James Letters.* 4 vols. Cambridge, Mass.: Harvard Univ. Press, 1974–84.

————. *The Life of Henry James.* 5 vols. Philadelphia and New York: J. B. Lippincott, 1952–72.

————, and Don H. Laurence, eds. *A Bibliography of Henry James.* London: Rupert Hart-Davis, 1961.

Edgerton, G. "Photography as One of the Fine Arts." *Craftsman* XII (July 1907).

Eichberg, Robert, and Jacqueline Quadow. Signal Corp Historical Project F-2b: *Combat Photography.* Washington, D.C.: Modern Military Branch, Military Archives Division, Nov. 1945.

Eisenstein, Sergei. *Film Essays and a Lecture.* Ed. Jay Leyda. New York: Praeger, 1970.

————. *Film Form.* New York: Harcourt, Brace and World, 1949.

Elias, Robert. *Theodore Dreiser: Apostle of Nature.* New York: Knopf, 1949.

Emerson, Peter Henry. *Naturalistic Photography for Students of the Art/The Death of Naturalistic Photography.* 1899; rpt. New York: Arno, 1973.

————. "Science and Art." In *Photographers on Photography.* Ed. Nathan Lyons. Englewood Cliffs, N.J.: Prentice-Hall & George Eastman, 1966.

Emerson, Ralph Waldo. *Journal.* 6 vols. Eds. Edward Waldo Emerson and Waldo Emerson Forbes. Boston: Houghton Mifflin, 1912.

————. "Nature." In *The Selected Writings of Ralph Waldo Emerson.* Ed. Brook Atkinson. New York: Modern Library, 1950.

Evans, Walker. *American Photographs.* 1938; rpt. New York: East River, 1975.

————. *Walker Evans at Work.* New York: Harper and Row, 1982.

"Exposure under Fire: An Official History of SigC Photography in the Luzon Operation." Washington, D.C.: U.S. Army SigC, S.W.P.A., 25 April 1945.

Falk, Stanley L. *Decision at Leyte.* New York: W. W. Norton, 1966.

Fallaci, Oriana. "An Interview with Norman Mailer." *Writers' Digest* (Dec. 1969): 40–47.

Faulkner, William. *A Fable*. New York: Random, 1950.

Fay, Brian. *Social Theory and Political Practice*. London: George Allen & Unwin, 1975.

Feldman, Seth R. *Dziga Vertov: A Guide to References and Resources*. Boston: G. K. Hall, 1979.

Fields, Annie. *Authors and Friends*. Boston: Houghton Mifflin, 1897.

Firebaugh, Joseph J. "Coburn: Henry James's Photographer." *American Quarterly* VII (Fall 1955): 215–33.

Fishman, J. *Jewish Radicals from Czarist Stetl to London Ghetto*. New York: Pantheon, 1974.

Fitzgibbon, John. "The 'Arkansaw Traveler' Daguerreotyped." *Photographic and Fine Art Journal* (Nov. 1854).

Foucault, Michel. *Discipline and Punish*. Trans. Alan Sheridan. New York: Vintage, 1979.

———. *The Order of Things: An Archaeology of the Human Sciences*. New York: Vintage, 1973.

———. *Power/Knowledge*. Trans. Colin Gordon *et al.* New York: Pantheon, 1980.

Frank, Robert. "Statement, 1958." In *Photography in Print*. Ed. Vicki Goldberg. New York: Simon and Schuster, 1981.

Frank, Waldo. "Alfred Stieglitz." *McCall's* (May 1927): 24–25.

———, *et al.*, eds. *America and Alfred Stieglitz: A Collective Portrait*. 1934; rpt. Millerton, N.Y.: Aperture, 1979.

———. *The Rediscovery of America*. New York: Scribner's, 1929.

Freund, Gisèle. *Photography and Society*. Boston: David R. Godine, 1980.

Fried, A., and R. M. Etman, eds. *Charles Booth's London: A Portrait of the Poor at the Turn of the Century, Drawn from His "Life and Labour of the People in London."* New York: Pantheon, 1968.

Furner, Mary O. *Advocacy and Objectivity: A Crisis in the Professionalization of American Social Science 1865–1905*. Lexington: Univ. of Kentucky Press, 1975.

Gadamer, Hans-Georg. *Philosophical Hermeneutics*. Trans. David Linge. Berkeley: Univ. of California Press, 1976.

Gaudin, Marc Antoine. *Traité pratique de photographie*. Paris: J. J. Dubochet, 1844.

Girard, René. *Deceit, Desire and the Novel*. Trans. Yvonne Freccero. Baltimore: Johns Hopkins Univ. Press, 1965.

Goddard, George. Overview: *A Life-long Adventure in Aerial Photography*. Garden City, N.Y.: Doubleday, 1969.

Goffman, Erving. *Encounters*. Indianapolis: Bobbs-Merrill, 1961.

———. *Interaction Ritual*. New York: Pantheon, 1967.

Goldberg, Vicki, ed. *Photography in Print: Writings from 1816 to the Present*. N.Y.: Simon and Schuster/Touchstone, 1981.

Goldschmidt, Lucien, and Weston J. Naef. *The Truthful Lens: A Survey of the Photographically Illustrated Book 1844–1914*. New York: Grolier Club, 1980.

Gollin, Rita K. *Portraits of Nathaniel Hawthorne*. DeKalb, Ill.: Northern Illinois Univ. Press, 1983.

Gouraud, Francois. *Description of the Daguerreotype Process*. Boston: Dutton and Wentworth, 1840.

Gray, J. Glenn. *The Warriors: Reflections on Men in Battle*. New York: Harcourt, Brace, 1959.

Green, Jonathan, ed. *Camera Work: A Critical Anthology*. Millerton, N.Y.: Aperture, 1973.

Greenfield, Kent R. *Command Decisions*. Washington, D.C.: Office of the Chief of Military History, United States Army, 1960.

"A Growl for the Unpicturesque." *Atlantic Monthly* XCVIII (1906): 140–43.

Guest, Antony. *Art and the Camera*. London, 1907; rpt. New York: Arno, 1973.

———. "Mr. Coburn's Vortographs." *Photo Era* XXXVIII (May 1917).

Habermas, Jürgen. *Knowledge and Human Interest*. Trans. Jeremy J. Shapiro. Boston: Beacon, 1971.

———. *Legitimation Crisis*. Trans. Thomas McCarthy. Boston: Beacon, 1975.

———. *Toward a Rational Society*. Trans. Jeremy Shapiro. Boston: Beacon, 1970.

Haines, Robert. "Alfred Stieglitz and the New Order of Consciousness in American Literature." *Pacific Coast Philology* VI (April 1971).

Hartmann, Sadakichi. "Alvin Langdon Coburn, Secession Portraiture." In *The Valiant Knights of Daguerre*. Ed. Harry W. Lawton and George Knox. Berkeley and Los Angeles: Univ. of California Press, 1978.

———. "The Broken Plates." *Camera Work* VI (April 1904).

———. "A New Departure in Photography." *The Lamp* XXVII (Feb. 1904).

Hawthorne, Nathaniel. *The Blithedale Romance*. Columbus: Ohio State Univ. Press, 1970.

———. *The House of the Seven Gables*. Columbus: Ohio State Univ. Press, 1970.

———. *Our Old Home*. Columbus: Ohio State Univ. Press, 1970.

———. *Transformation or the Romance of Monte Beni*. 2 vols. Leipzig: Bernhard Tauchnitz, 1860.

———. *Twice Told Tales*. New York: E. P. Dutton, 1961.

Heller, Erich. "Literature and Political Responsibility." *Commentary* (July 1971.)

Hellman, Lillian. *An Unfinished Woman*. Boston: Little, Brown, 1969.

Herbst, Josephine. "The Starched Blue Sky of Spain." *The Noble Savage* I (1960).

Higgins, Charles. "Photographic Aperture: Coburn's Frontispieces to James's New York Edition." *American Literature* 53 (1982): 661–75.

Higginson, T. W. *Letters and Journals of T. W. Higginson, 1846–1906*. Ed. M. T. Higginson. Boston: 1921; rpt. New York: Negro Univ. Press, 1969.

Hill, Paul, and Thomas Cooper. *Dialogue with Photography*. New York: Farrar, Straus and Giroux, 1979.

Himmelfarb, Gertrude. *The Idea of Poverty*. New York: Knopf, 1984.

Hodge, Peyton, Thomas Hooper and Victor Lott. *Avengers of Bataan: 38th Infantry Division, Luzon Campaign*. Atlanta: Albert Lore, 1947.

Hofstader, Richard. *The Age of Reform*. New York: Knopf, 1959.

Hollowell, John. *Fact and Fiction: The New Journalism and the Nonfiction Novel*. Chapel Hill: Univ. of North Carolina Press, 1977.

Holmes, Oliver Wendell. "Doings of the Sunbeam." *Atlantic Monthly* XII (1863): 1–15.

———. "The Stereoscope and the Stereograph." *Atlantic Monthly* III (June 1859): 738–48.

———. "Sun Painting and Sun Sculpture." *Atlantic Monthly* VIII (1861): 13–29.

Homer, William Innes. *Alfred Stieglitz and the American Avant-garde*. Boston: New York Graphic Society, 1977.

———. *Alfred Stieglitz and the Photo-Secession*. Boston: Little, Brown, 1983.

Horkheimer, Max, and Theodor Adorno. *Dialectics of Enlightenment*. New York: Seabury, 1972.

Howard, Clive, and Joe Whitley. *One Damned Island after Another*. Chapel Hill: Univ. of North Carolina Press, 1946.

Howells, William Dean. *A Hazard of New Fortunes*. New York: Harper and Bros., 1890.

———. *London Films*. London: Harper and Bros., 1906.

Hoyle, Pamela. *The Development of Photography in Boston, 1840-1875*. Boston: Boston Athenaeum, 1978.

Hunt, Robert. *The Poetry of Science, or Studies of Physical Phenomena of Nature*. Boston: Gould, Kendall and Lincoln, 1950.

———. *A Popular Treatise on the Art of Photography, inc. Daguerreotype, and All the New Methods of Producing Pictures by the Chemical Agency of Light*. Glasgow: Richard Griffin, 1841.

Hunter, Robert. *Poverty*. New York: Macmillan, 1904.

Hurley, F. J. *Portrait of a Decade: Roy Stryker and the Development of Documentary Photography in the Thirties*. Rochester, N.Y.: Light Impressions, 1972.

Hyde, Lewis. *The Gift: Imagination and the Erotic Life of Property*. New York: Vintage, 1983.

Images of America: Early Photography: 1839-1900. Washington, D.C.: Library of Congress, 1957.

Ivens, Joris. *The Camera and I*. New York: International Publishers, 1969.

James, Henry. *The American Scene*. New York: Horizon, 1967.

———. *The Art of the Novel: Critical Prefaces*. Ed. R. P. Blackmur. New York: Scribner's, 1934.

———. *Future of the Novel*. Ed. Leon Edel. New York: Vintage, 1956.

———. *The Novels and Tales of Henry James: New York Edition*. 26 vols. New York: Scribner's, 1907-9; 1917.

———. *The Painter's Eye: Notes and Essays on the Pictorial Arts*. Ed. John L. Sweeney. Cambridge, Mass.: Harvard Univ. Press, 1956.

———. *Picture and Text*. New York: Harper and Bros., 1893.

Jameson, Fredric. "Marxism and Historicism." *New Literary History* 11 (Autumn 1979): 41-73.

———. *The Political Unconscious*. Ithaca, N.Y.: Cornell Univ. Press, 1981.

———. *The Prison House of Language*. Princeton: Princeton Univ. Press, 1972.

Jamieson, Stuart, *Labor Unionism in American Agriculture*. Washington, D.C.: Bulletin No. 836 of the Bureau of Labor Statistics, 1945.

Janeway, Elizabeth. *Powers of the Weak*. New York: William Morrow, 1981.

Jay, Martin. *The Dialectical Imagination. A History of the Frankfurt School and the Institute of Social Research*. Boston: Little, Brown, 1973.

Jenkins, Reese V. "Technology and the Market: George Eastman and the Origins of Mass Amateur Photography." *Technology and Culture* XVI (1975): 1-19.

Jones, Gareth Stedman. *Outcast London*. Oxford: Clarendon, 1971.

Jones, James. *The Thin Red Line*. New York: Scribner's, 1962.

Jaswick, Jerry, and Lawrence Keating. *Combat Cameraman*. Philadelphia: Chilton, 1961.

Joyce, James. *Stephen Hero*. New York: New Directions, 1963.

Jussim, Estelle. "Icons or Ideology: Stieglitz and Hine." *Massachusetts Review* (Winter 1978).

———. *Visual Communication and the Graphic Arts: Photographic Technologies in the Nineteenth Century*. New York: R. R. Bowker, 1974.

Kael, Pauline. "Celebrities Make Spectacles of Themselves." *New Yorker* XX (Jan. 1968): 90-95.

Kehl, D. G. "Steinbeck's 'String of Pictures' in *The Grapes of Wrath.*" *Image* XVII (March 1974).

Klaue, Wolfgang, Hans Wegner, and Manfred Lichtenstein. *Joris Ivens: Zusammenstellung und Redaktion.* Berlin: Staatlichen Filmarchiv der Republik, 1963.

Kolakowski, Leszek. *Main Currents of Marxism: The Breakdown.* New York: Oxford Univ. Press, 1981.

Kortian, Garbis. *Metacritique. The Philosophical Argument of Jürgen Habermas.* Trans. John Raffan. Cambridge: Cambridge Univ. Press, 1980.

Kozloff, Max. "Critical and Historical Problems of Photography." In *Renderings: Critical Essays on a Century of Modern Art.* New York: Simon and Schuster/Clarion, 1969.

Krook, Dorothea. *The Ordeal of Consciousness.* Cambridge: Cambridge Univ. Press, 1962.

Lange, Dorothea. *Dorothea Lange Farm Security Administration Photographs: 1935-1939.* 2 vols. Eds. Howard M. Levin and Katherine Northrup. Glencoe, Ill.: Text-Fiche, 1980.

————, and Paul Schuster Taylor. *An American Exodus: A Record of Human Erosion.* 1939; rpt. New Haven: Yale Univ. Press, 1969.

Langer, Elinor, *Josephine Herbst: The Story She Could Never Tell.* Boston: Little, Brown, 1984.

Lanser, Susan Sniader. *The Narrative Act: Point of View in Prose Fiction.* Princeton: Princeton Univ. Press, 1981.

Lathrop, George P. *A Study of Hawthorne.* Boston: James R. Osgood, 1876.

Lease, B. "Diorama and Dream: Hawthorne's Cinematic Vision." *Journal of Popular Culture* V (1971).

Le Carré, John. "Introduction." In *Hearts of Darkness: Photographs by Don McCullin.* New York: Knopf, 1981.

Leonard, Thomas. *Above the Battle, War-making in America from Appomattox to Versailles.* New York: Oxford Univ. Press, 1978.

Leyda, Jay. *Kino: A History of the Russian and Soviet Film.* New York: Collier, 1973.

London, Jack. *The People of the Abyss.* London: Macmillan, 1903.

Lowe, Sue Davidson. *Stieglitz.* New York: Farrar, Straus and Giroux, 1983.

Lowry, E. D. "The Lively Art of *Manhattan Transfer.*" *Publication of the Modern Language Assoc.* LXXIV (Oct. 1969): 1628-38.

Ludington, Townsend. *John Dos Passos: A Twentieth Century Odyssey.* New York: E. P. Dutton, 1980.

Lukacs, Georg. *Realism in Our Time.* New York: Harper and Row, 1962.

————. *The Theory of the Novel,* Trans. Anna Bostock. Cambridge: M.I.T. Press, 1971.

Magny, Claude-Edmonde. *The Age of the American Novel: The Film Aesthetics of Fiction between the Two Wars.* Trans. Eleanor Hochman. New York: Frederick Ungar, 1972.

Maholy-Nagy, Lázló. "From Pigment to Light." In *Photographers on Photography.* Ed. Nathan Lyons. Englewood Cliffs, N.J.: Prentice-Hall, 1966.

Mailer, Norman. *Armies of the Night.* New York: New American Library, 1971.

————. *The Deer Park.* New York: Putnam Publishing Group, 1981.

————. "First Advertisement for Myself." "The White Negro." In *The Long Patrol: 25 Years of Writing from the World of Norman Mailer.* Ed. Robert F. Lucid. New York: World Publishers, 1971.

————. *The Naked and the Dead.* New York: Rinehart, 1948.

——. "Some Dirt in the Talk: A Candid History of an Existential Movie Called *Wild 90.*" *Esquire* LXVIII (Dec. 1967).

"Mr. Mailer Interviews Himself." *New York Times Book Review* (17 Sept. 1967): 4-5.

Malcolm, Janet. *Diana and Nikon.* Boston: David R. Godine, 1980.

Marcuse, Herbert. *Negations.* Trans. Jeremy Shapiro. Boston: Beacon, 1968.

Matthiessen, F. O. *American Renaissance.* New York: Oxford Univ. Press, 1941.

——. *Theodore Dreiser.* New York: William Sloan Assoc., 1951.

——, and Kenneth B. Murdock, eds. *The Notebooks of Henry James.* New York: Oxford Univ. Press, 1947.

Mayhew, Henry. *London Labor and the London Poor.* 4 vols. London: Griffin, Bohn, 1861–62.

Mayne, J. *The Ideologies of Metacinema.* Ann Arbor, Mich.: Univ. Microfilms, 1975.

McCarthy, Thomas. *The Critical Theory of Jürgen Habermas.* Cambridge: M.I.T. Press, 1978.

McCullin, Don. *Hearts of Darkness.* New York: Knopf, 1981.

McWilliams, Carey. *Factories in the Field: The Story of Migratory Farm Labor in California.* Boston: Little, Brown, 1939.

Mehlman, Jeffrey. *Revolution and Repetition: Marx/Hugo/Balzac.* Berkeley and Los Angeles: Univ. of California Press, 1977.

Meltzer, Milton. *Dorothea Lange: A Photographer's Life.* New York: Farrar, Straus and Giroux, 1978.

Members of the National Committee for the Defense of Political Prisoners. *Harlan Miners Speak: Report on Terrorism in the Kentucky Coal Fields.* New York: Harcourt, Brace, 1932.

Mitchell, W. J. T., ed. *The Language of Images.* Chicago: Univ. of Chicago Press, 1974.

Moers, Ellen. *Two Dreisers.* New York: Viking, 1969.

Millgate, Michael. *American Social Fiction: James to Cozzens.* New York: Barnes and Noble, 1965.

Moreau, Geneviève. *The Restless Journey of James Agee.* New York: William Morrow, 1977.

Moyes, Norman Barr. "Major Photographers and the Development of Still Photography in Major American Wars." Thesis. Ann Arbor, Mich.: Univ. Microfilms, 1967.

Mydans, Carl. *More than Meets the Eye.* New York: Harper and Bros., 1959.

Mydans, Shelley Smith. *The Open City.* Garden City, N.Y.: Doubleday, Doran, 1945.

"The New Art Daguerreotype." *New York Morning Herald* (4 Oct. 1839).

Newhall, Beaumont. *The Daguerreotype in America.* New York: Duell, Sloan and Pearce, 1961.

Nochlin, Linda. "The Realist Criminal and the Abstract Law II." *Art in America* LXI (Nov.–Dec. 1973).

Norman, Dorothy, ed. "From the Writings and Conversations of Alfred Stieglitz." *Twice-a-Year* I (Fall–Winter 1938).

"Norman Mailer." *Current Biography* (Oct. 1948).

Nye, Russell B. "Photography and American Culture." In *Toward a New American Literary History.* Eds. Louis J. Budd, Edwin H. Cady, and Carl L. Anderson. Durham, N.C.: Duke Univ. Press, 1980.

Ohlin, P. H. *Agee.* New York: Ivans Obolensky, 1966.

Ohmann, Richard. "Politics and Genre in Nonfiction Prose." *New Literary History* (1980): 237–44.

Ohrn, Karin B. *Dorothea Lange and the Documentary Tradition.* Baton Rouge: Louisiana State Univ. Press, 1980.

Owens, Bill. *Documentary Photography: A Personal View.* Danbury, N.H.: Addison House, 1978.

Parrington, V. L. *The Beginnings of Critical Realism in America.* New York: Harcourt, Brace, 1930.

Parry, Albert. *Garrets and Pretenders: A History of Bohemianism in America.* New York: Covici-Friede, 1933.

Petrey, Sandy. "The Language of Realism, The Language of False Consciousness: A Reading of *Sister Carrie.*" *Novel* IX (Winter 1977): 101–13.

Pizer, Donald. *Novels of Theodore Dreiser.* Minneapolis: Univ. of Minnesota Press, 1976.

Powers, Lyall. *Henry James and the Naturalist Movement.* Michigan: Michigan State Univ. Press, 1971.

Praz, Mario. *Mnemosyne: The Parallel between Literature and the Visual Arts.* Princeton: Princeton Univ. Press, 1967.

Prime, Samuel I. *The Life of Samuel Morse.* New York: Appleton, 1975

Raines, Howell. "Let Us Now Revisit Famous Folk." *New York Times Magazine* (25 May 1980).

Regler, Gustav. *The Owl of Minerva.* Trans. Norman Denny. New York: Farrar, Straus and Cudahy, 1959.

Riess, Suzanne. "Dorothea Lange: The Making of a Documentary Photographer." Interview. Berkeley: Univ. of California Regional Oral History Office, 1968.

Riis, Jacob A. *The Battle with the Slum.* New York: Macmillan, 1902.

———. *How the Other Half Lives.* 1896; rpt. New York: Dover, 1971.

———. *The Making of an American.* New York: Macmillan, 1904.

———. *A Ten Years War.* Boston: Houghton Mifflin, 1900.

———. *Theodore Roosevelt, The Citizen.* New York: Outlook, 1904.

Robinson, Henry Peach. *The Elements of a Pictorial Photograph.* 1896; rpt. New York: Arno, 1973.

———. *Pictorial Effect in Photography.* New York: Scovill and Adams, 1897.

Root, Marcus, A. *The Camera and the Pencil.* Philadelphia: H. B. Lippincott, 1864.

Rosen, Robert C. *John Dos Passos: Politics and the Writer.* Lincoln: Univ. of Nebraska Press, 1981.

Rosenfeld, Paul. *Port of New York.* New York: Harcourt, Brace, 1924.

Rosenblum, Barbara. *Photographers at Work: A Sociology of Photographic Style.* New York: Holmes and Meiser, 1978.

Rousseau, Jean Jacques. *The Social Contract.* London: Swan Sonnenschein, 1898.

Rowe, John Carlos. *The Theoretical Dimensions of Henry James.* Madison: Univ. of Wisconsin Press, 1984.

———. *Through the Custom-House: Nineteenth-century American Fiction and Modern Theory.* Baltimore: Johns Hopkins Univ. Press, 1982.

Rudisill, Richard. *Mirror Image: The Influence of the Daguerreotype on American Society.* Albuquerque: Univ. of New Mexico Press, 1971.

Ryder, James F. *Voightlander and I in Pursuit of Shadow Catching.* Cleveland: Cleveland Printing and Publishing, 1902.

Scharf, Aaron. *Art and Photography.* Harmondsworth: Penguin, 1968.

———. *Pioneers of Photography.* New York: Harry N. Abrams, 1975.

Schiller, Dan. "Realism, Photography, and Journalistic Objectivity in 19th Century America." *Studies in the Anthropology of Visual Communication* IV (1977): 86–98.

Schor, Naomi. "Zola: From Window to Window." *Yale French Studies* XLII (1969).

Schuneman, R. S., ed. *Photographic Communication: Principles, Problems and Challenges to Photo Journalism.* Rochester, N. Y.: Light Impressions, 1972.

Segal, Ora. *The Lucid Reflector: The Observer in Henry James' Fiction.* New Haven: Yale Univ. Press, 1969.

Sekula, Allan. "Reinventing Documentary." In *Photography and Language.* Ed. Lew Thomas. San Francisco: Camerawork, 1976.

Seligmann, Herbert. *Alfred Stieglitz Talking.* New Haven: Yale Univ. Library, 1966.

Seltzer, Mark. *Henry James and the Art of Power.* Ithaca, N.Y.: Cornell Univ. Press, 1984.

———. "The Princess Casamassima: Realism and the Fantasy of Surveillance." In *American Realism: New Essays.* Ed. Eric J. Sundquist. Baltimore: Johns Hopkins Univ. Press, 1982.

Sennett, Richard and Jonathan Cobb. *The Hidden Injuries of Class.* New York: Vintage, 1973.

Seventh Special Report of the Commissioner of Labor. "The Slums of Baltimore, Chicago, New York and Philadelphia." Washington, D.C.: G.P.O., 1894.

Sherman, Cindy. *Cindy Sherman.* Ed. Peter Schjeldahl. New York: Pantheon, 1984.

Sherman, Stuart P. "The Naturalism of Mr. Dreiser." In *On Contemporary Literature.* New York: Holt, Rinehart and Winston, 1917.

Shloss, Carol. "Oliver Wendell Holmes as an Amateur Photographer." *History of Photography* V (April 1981).

Shockly, Martin Staples. "The Reception of *The Grapes of Wrath* in Oklahoma." *American Literature* XV (May 1944): 351–61.

Sims, George. *The Mysteries of Modern London.* London: C. Arthur Pearson, 1906.

Smith, Adolphe. *Street Life in London with Photographs by John Thompson.* London: 1877; rpt. Bronx, New York: Benjamin Blum, 1969.

Smith, Robert Ross. *The War in the Pacific: Triumph in the Philippines.* United States Army in World War II Series. Washington, D.C.: Office of the Chief of Military History, United States Army, 1963.

Snyder, Joel. "Picturing Vision." *Critical Inquiry* VI (1980): 499–526.

Snyder, Joel, and Neil Walsh Allen. "Photography, Vision and Representation." *Critical Inquiry* (Autumn, 1975).

Sobieszek, Robert. *The Spirit of Fact: The Daguerreotypes of Southworth and Hawes: 1843–1862.* Rochester: International Museum of Photography, 1976.

Society for Promoting Christian Knowledge. *The Wonders of Light and Shadow.* 1851; rpt. New York: Arno, 1973.

Solow, H. "Substitution at Left Tackle: Hemingway for Dos Passos." *Partisan Review* 4 (April 1938).

Sontag, Susan. *On Photography.* New York: Farrar, Straus and Giroux, 1977.

"Speaking of Pictures: These by *Life* Prove Facts in *Grapes of Wrath.*" *Life* (19 Feb. 1940): 10–11.

Spiegel, Alan. *Fiction and the Camera Eye.* Charlotte: Univ. of Virginia Press, 1976.

Stanley, Roy M. *World War II Photo Intelligence.* New York: Scribner's, 1981.

Samaras, Lucas. *Samaras Album.* New York: Whitney Museum of Art, 1971.

Stebbins, Robert, and Jay Leyda. "Joris Ivens: Artist in Documentary." *Magazine of Art* (July 1948).

Steichen, Edward. *A Life in Photography.* Garden City, N.Y.: Doubleday, 1963.

Stein, Sally. "Making Connections with the Camera: Photography and Social Mobility in the Career of Jacob Riis." *Afterimage* X (May 1983): 11–16.

Stein, Walter J. *California and the Dust Bowl Migration.* Westport, Conn.: Greenwood, 1973.

Steinbeck, John. *Bombs Away: The Story of a Bomber Team.* New York: Viking, 1942.

———. "Dubious Battle in California." *Nation* CXLIII (Sept. 1936): 302–4.

———. *The Grapes of Wrath.* New York: Viking, 1939.

———. *In Dubious Battle.* New York: Covici-Friede, 1936.

———. *Once There Was a War.* New York: Viking, 1958.

———. *Their Blood Is Strong.* San Francisco: Simon J. Lubin Society of California, 1938.

———. *A Russian Journal with Pictures by Robert Capa.* New York: Viking, 1948.

Stern, J. P. "Reflections on Realism." *Journal of European Studies* VII (March 1971).

Stieglitz, Alfred. "The Camera Workers." TS. Beinecke Library, Yale Univ.

———. "How I Came to Photograph Clouds." *The Amateur Photographers and Photography* LVI (1923).

———. "Modern Pictorial Photography." *The Century Illustrated Monthly Magazine* LXIII (Oct. 1902): 822–26.

"Alfred Stieglitz and His Latest Work." *The Photographic Times* XXVIII (April 1896).

"Alfred Stieglitz: Four Happenings." *Twice-a-Year* VII–IX (1942).

Stott, William. *Documentary Expression and Thirties America.* New York: Oxford Univ. Press, 1973.

———. "Walker Evans: A Memoir and Introduction." In *Walker Evans: Photographs from the "Let Us Now Praise Famous Men" Project.* Austin: Univ. of Texas Humanities Research Center, 1974.

Sundquist, Eric, ed. *American Realism: New Essays.* Baltimore: Johns Hopkins Univ. Press, 1982.

Swados, Harvey, ed. *The American Writer and the Great Depression.* New York: Bobbs-Merrill, 1966.

Swanberg, W. A. *Dreiser.* New York: Scribner's, 1965.

Taft, Robert. *Photography and the American Scene: A Social History: 1839–1889.* New York: Dover, 1964.

Talbot, William Henry Fox. *The Pencil of Nature.* London: 1844; rpt. in *Image* 2 (June 1959).

Taylor, Paul Schuster. *On the Ground in the Thirties.* Salt Lake City: Peregrine Smith, 1983.

———, and Clark Kerr. "Documentary History of the Strike of the Cotton Pickers in California, 1933." *Violations of Free Speech and Rights of Labor: Hearings Before a Subcommittee of the Committee on Education and Labor,* U.S. Senate, Seventy-sixth Congress, Third Session; Part 54, "Agriculture Labor in California." Washington, D.C.: G.P.O., 1940.

Terkel, Studs, ed. *Hard Times: An Oral History of the Great Depression.* New York: 1970.

Theroux, Paul. *Picture Palace.* New York: Ballantine, 1978.

13,000 Hours: Combat History of the 32nd Infantry Division, World War II. Prepared by the Public Relations Office, 32nd Inf. Div. under the Direction of AC of S, G-Z. Manila, 1945.

Thomas, Alan. *Time in a Frame: Photography and the Nineteenth Century Mind.* New York: Schocken, 1977.

Thompson, G. R., and Dixie R. Harris. *The Technical Services: The Signal Corps: The Outcome.* United States Army in World War II Series. Washington, D.C.: Office of the Chief of Military History, United States Army, 1966.

Thompson, Jerry L. *Walker Evans at Work*. New York: Harper and Row, 1982.

Thoreau, Henry David. *Journal*. 2 vols. Ed. Bradford Torrey. Boston: Houghton Mifflin, 1912.

Trabold, Edward R. "Counterattacking with a Camera." *Photographic Journal of America* CVI (1919): 407–11.

Trachtenberg, Alan, ed. *Classic Essays on Photography*. New Haven: Leete's Island, 1980.

———. *The Incorporation of America: Culture and Society in the Gilded Age*. New York: Hill and Wang, 1982.

———. "Lewis Hine: The World of His Art." In *Photography in Print: Writings from 1816 to the Present*. Ed. Vicki Goldberg. New York: Simon and Schuster/Touchstone, 1981.

U.S. Army, Navy and Air Force. *Photographic Interpreter's Handbook*. Washington, D.C.: 1954.

U.S. Army Technical Manual: TM 11-401-2. "Army Pictorial Techniques, Equipments and Systems, Still Photography." Washington, D.C.: G.P.O.

U.S. Army Technical Manual: TM 30-245. "Image Interpretation Handbook." Washington, D.C.: G.P.O.

U.S. Defense Intelligence Agency Publications: DIAM 55-5. "Aerial Photography and Airborne Electronic Sensor Imagery."

U.S. War Department. *Advanced Map and Aerial Photography Reading: FA 21-26*. Washington, D.C.: G.P.O., 1941.

Valtin, Jan [pseud.]. *Children of Yesterday*. New York: Readers', 1946.

Vertov, Dziga. "The Man with the Movie Camera." Trans. Vladimir Petric. TS. Fogg Art Museum Library.

———. "The Vertov Papers." Trans. Chelovek and Konapparatom. *Film Comment* VIII (Spring 1972): 46–51.

———. "The Writings of Dziga Vertov." In *Film Culture Reader*. Ed. P. A. Sitney. New York: Praeger, 1970.

Wagner, Linda, *Dos Passos: Artist as American*. Austin: Univ. of Texas Press, 1979.

Walcutt, Charles Child. *American Literary Naturalism: A Divided Stream*. Minneapolis: Univ. of Minnesota Press, 1956.

Ware, Louise. *Jacob A. Riis: Police Reporter, Reformer, Useful Citizen*. New York: Appleton-Century, 1938.

Wegner, Hans. *Joris Ivens: Dokumentarist der Wahrheit*. Berlin: Henschelverlag Kunst und Gesellschaft, 1965.

Weimann, Robert. *Structure and Society in Literary History*. Charlottesville: Univ. of Virgina Press, 1976.

Weinstein, Philip M. *Henry James and the Requirements of the Imagination*. Cambridge, Mass.: Harvard Univ. Press, 1971.

Welty, Eudora. *One Writer's Beginnings*. Cambridge, Mass.: Harvard Univ. Press, 1984.

Westerhaven, James N. "Autobiographical Elements in the Camera Eye." *American Literature* XLVII (Nov. 1948): 340–64.

Weston, Edward. *The Daybooks of Edward Weston*. Ed. Nancy Newhall. Rochester: George Eastman House, 1961.

Wharton, Edith. *The Age of Innocence*. New York: Scribner's, 1968.

Whelan, Richard. *Robert Capa: A Biography*. New York: Knopf, 1985.

Wiebe, Robert H. *The Search for Order: 1877–1920*. New York: Hill and Wang, 1966.

Williams, Raymond. *Marxism and Literature*. London: Oxford Univ. Press, 1977.

Williams, William Carlos. *Paterson*. New York: New Directions, 1963.

Willis, Nathaniel P. "The Pencil of Nature." *The Corsair* (April 1839): 71.

Wilscher, Ann. "Photography in Literature." *History of Photography* II (July 1978).

Wouk, Herman. *The Cane Mutiny*. Garden City, N.Y.: Doubleday, 1951.

Wright, Major B. C. *The First Cavalry in World War II*. Tokyo: Toppan, 1947.

Wuthnow, Robert, James Davidson Hunter, Albert Bergesen, and Edith Kurzweill. *Cultural Analysis: The Works of Peter L. Berger, Mary Douglas, Michel Foucault and Jürgen Habermas*. Boston: Routledge & Kegan Paul, 1984.

Youmans, Edward L. "Under False Colors." *Popular Science Monthly* VII (July 1875).

Ziolkowski, T. *Disenchanted Images: A Literary Iconology*. Princeton: Princeton Univ. Press, 1977.

INDEX

Adams, Ansel, 188
Adams, Henry, 96
Adams, John Quincy, 34
Addams, Jane, 79
Adorno, Theodor, 6, 274, 281
Aerial surveillance, 234–237
Agee, James: 7, 8, 14, 19, 179–197, 211, 255; cricitism of Margaret Bourke-White, 181–187, 190; and Walker Evans, 179–197; and Farm Security Administration, 179, 189; as film critic, 179; and *Fortune* Magazine, 179, 180, 186, 193; on importance of the camera, 14, 15; *Let Us Now Praise Famous Men*, 180, 186, 197; as observer, 193–196; on trust, 196, 197
American Scene, The, 93–96
American Social Science Association, 119–120
Anderson, Sherwood, 99
Antin, Mary, 101
Arbus, Diane, 10–14, 247–248, 274
Army Signal Corp, 240
Atlantic Monthly, 25–27, 37, 38

Bakhtin, M. M., 267
Balzac, Honoré de, 65
Barnum, P. T., 30
Beckett, Samuel, 14, 260
Benson, Charles, 224
Bentham, Jeremy, 13, 14, 210
Berger, John: 5, 6, 11, 12, 19, 241, 262; on French peasants, 5, 6, 115; *Into Their Labors*, 4–6; on Paul Strand, 12, 203; on photography as memory, 264–265; on primitive art, 117
Blake, William, 103
Blanchot, Maurice, 253
Blithedale Romance, The, 45–47, 74, 75
Booth, Charles, 121
Booth, General William, 87, 88
Boston, 30–31, 37
Bourke-White, Margaret: 20, 190, 192, 196–197, 207, 236; Agee on, 181–185, 190; on

photography, 183; and photo journalism, 182; trip to South, 182–185; *You Have Seen Their Faces*, 181–186
Bowen, Elizabeth, 110
Brace, Charles Loring, 95, 119, 121–122, 128, 129
Brady, Mathew, 38
Bringing in the Sheaves, 222, 226, 227
"Broken Plates," 58
Brook Farm, 45
Brooklyn Bridge, 99, 103
Burns, Robert, 25

Caffin, Charles, on Matisse, 105
Caldwell, Erskine, 182
California State Emergency Relief Administration, 204, 205, 206, 224
Camera: approaching with, 6–14, 21; Agee on, 7–8; Arbus on, 10, 13, 14; as art, 4, 5, 56, 63–64, 189; Cartier-Bresson on, 7–8; Evans on, 7–8; invention of, 33; as language of neutrality, 17; McCullin on, 9–10; and nature of photography, 66; Stieglitz on, 8–9; Weston on, 8, 9. *See also* Daguerreotypy
Camera Work, 55
Cantor, Jay, 21, 256
Capa, Robert, 20, 21, 241, 247
Caputo, Philip, 256–258
Cartier-Bresson, Henri, 7–11, 239, 274
Chatfield-Taylor, Hobart, 116
Chesterton, Mrs. Cecil, 78–88
Clark, Anson, 33, 34
Coates, Joseph, 116, 117
Coburn, Alvin Langdon: 55–89; and *Century Magazine*, 59; and composition, 63; Sadakichi Hartmann on, 62; and instruction from James, 68–69; and James, 59–71; and landscape photography, 63, 70; and London, 61, 67–68, 82; *Men of Mark* series, 70; photography as art, 63–64; and Photo Secession group, 62; portraits, 70; and teachers, 62

303